Graphical Communication

A three-book course intended for students preparing for examinations in Graphical Communication and Technical Drawing at the age of sixteen plus.

The authors:

B. G. CUTHBERT

Deputy Head, Diss High School, Norfolk.

Chief Examiner:

Geometrical and Building Drawing,
East Anglian Examinations Board, 1966-1973

Technical Studies — Drawing,
North Regional Examinations Board, 1970-1977

Graphical Communication (Technical Drawing)
Metropolitan Regional Examinations Board, 1978-1979

Technical Drawing (Engineering) G.C.E. 'O' Level,
University of London, 1974-1980

Graphical Communication G.C.E. 'O' Level,
University of London, 1981-1986

Technical Drawing
The West Midlands Examinations Board, 1986

Engineering Drawing and Building Drawing G.C.E. 'O' Level,
Welsh Joint Education Committee, 1986

M. R. PATTENDEN

Deputy Head, Lealands High School, Luton.

Chief Examiner:

Geometrical and Engineering Drawing,
East Anglian Examinations Board, 1968-1983

Preface

This is the second of two volumes intended for students preparing for all examinations in Graphical Communication and Technical Drawing at the age of sixteen plus.

The subject matter contained in these volumes satisfies the requirements of current CSE and GCE 'O' level syllabuses in Graphical Communication and Technical Drawing in their widest sense and includes logos, symbols, ideograms, graphs and charts, flow diagrams, instruction sequences, maps and layouts of roads and buildings etc., in addition to drafting methods of graphical illustration and applied geometry. Areas of study also included are engineering drawing, school creative arts and electrical circuits.

Considerable emphasis in each book is given to elements of graphical and structural design.

Each section contains examples intended to extend the capabilities of the most able pupils in addition to making the necessary provision for those requiring repetition at a less high level. Examples of work are taken from the everyday experiences of the pupils.

Both authors have considerable experience of examining Graphical Communication and Technical Drawing in all its aspects over a considerable number of years, and they have prepared this series of books based on that experience.

Book 2 is intended particularly for 4th and 5th year pupils engaged on an examination course in Graphical Communication and Technical Drawing and includes all material for CSE and core material for GCE and 16+ examinations.

Acknowledgements

The authors gratefully acknowledge permission to reproduce material as follows:
Motor Cycle News and Honda (UK) Ltd, front cover;
J. A. Bacchelor Fig. 4.1; Reckitt & Colman 7.15; Goblin Ltd. 7.16; Dunlop 7.26; Letraset UK 8.1, 8.2, 8.3; Townsend Thoresen 9.2; Ofrex Ltd 10.11 to 10.14 and 10.16 to 10.18 inclusive; Rotring UK 10.18; Winsor & Newton 10.19; Mazda Cars (UK) Ltd 10.23; Berol Limited 10.24; TI Group plc 11.5, 11.7; J. A. Bacchelor 12.1, 12.2, 12.3; Honda (UK) Ltd 12.7; Space Frontiers Limited 14.13.

© 1984 B.G. Cuthbert and M.R. Pattenden
0 00 322051 6

Published by Collins Educational
8 Grafton Street, London W1X 3LA

First impression 1984
Reprinted 1985

2 3 4 5 6 7 8 9

Printed in Great Britain at the
University Press, Cambridge

All rights reserved. No part of this publication may be reproduced, stored in a retrieval system, or transmitted in any form or by any means, electronic, mechanical, photocopying, recording or otherwise, without prior permission of the copyright owners.

Contents

		Page
1	Freehand drawing	7
	Freehand drawing techniques	10
2	Drafting methods	13
	Sequence drawings	13
	Exploded views	15
3	Applied geometry 1	17
	Division of a line	17
	Construction of angles	18
	Construction of regular polygons	20
4	Electrical circuits	24
5	Applied geometry 2	26
	Enlargement and reduction	26
6	Charts	28
	Flow charts	28
	Time charts	30
7	Applied geometry 3	32
	Circles	32
	Tangents	35
	Tangential arcs	37
8	The use of dry-transfer method of lettering, shading and toning	40
9	Applied geometry 4	41
	The ellipse	41
10	The use of shading and colour	47
11	Graphs	49
12	Applied geometry 5	51
	Loci	51
	The cycloid	56
	The helix	57
13	Applied geometry 6	59
	Development – prisms	59
	– cylinders	61
	– pyramids	63
	– cones	65

		Page
14	Projects	67
	Tractor project	67
	Lantern project	69
	Gemini project	70
	Veteran car project	71
	Toy roller project	73
	Card modeller's clamp project	74
15	Applied geometry 7	75
	Intersection – prisms	75
	– cylinders	77
16	Pictorial projection	79
	Oblique	79
	Isometric	80
17	Dimensioning	84
	Dimensioning guide	85
18	Sections	87
19	Screw threads	89
20	Engineering graphics	91

Graphical Communication

Book 2

B. G. Cuthbert
Deputy Head, Diss High School, Norfolk

M. R. Pattenden
Deputy Head, Lealands High School, Luton

COLLINS EDUCATIONAL   LONDON AND GLASGOW

1 Freehand drawing

One important aspect of graphical communication is the representation in two-dimensional form on a flat sheet of paper of solid, three dimensional objects, such as the part of a door stop shown in the photograph Fig. 1.1.

This can be done either by using instruments or by drawing freehand. The same principles and techniques are as important in freehand drawing as in instrument work.

It is essential that the object drawn should be recognised easily. Producing well-proportioned and easily recognisable freehand drawings comes only after practice. Some techniques which can help you develop your drawing are given on the following pages. One particular aid is the use of grid paper and Figs 1.2 and 1.3 show pictorial and orthographic freehand views of the door stop seen in the photograph.

Note the quality of line that is acceptable for freehand drawing.

When you practise freehand drawing, try always to work from the object itself rather than from another drawing of it. Then, the materials used to construct the object are taken into consideration. The material used can affect the shape and the mode of construction of the object.

Fig. 1.1

Fig. 1.2 **Fig. 1.3**

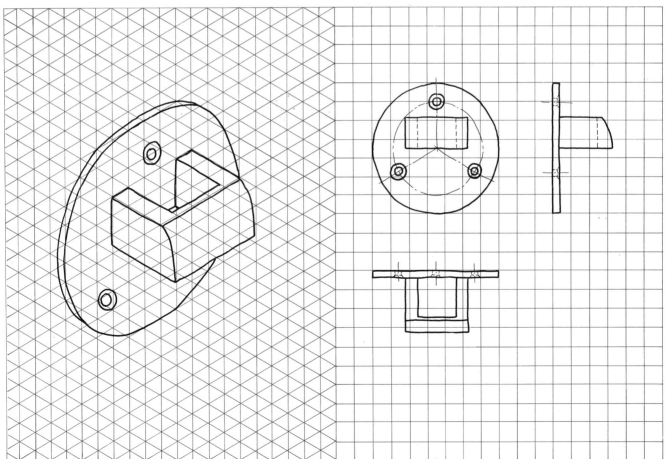

8 Freehand drawing

Fig. 1.4

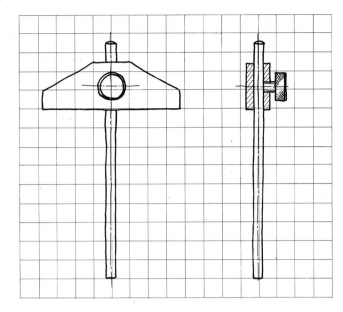

When drawing an orthographic view on grid paper, use the grid lines to give guidance for vertical and horizontal lines and for size. The squares on grid paper usually have sides of 5 or 10 mm; by counting the squares, a drawing of the correct proportions can be produced easily.

Note that the conventions used in instrument work for centre lines, section shading, etc. have been used in the freehand views of the depth gauge (Fig. 1.4). Note how the materials used in the construction of the depth gauge have affected the proportions of the tool.

Fig. 1.5

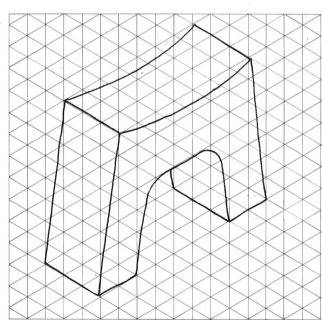

When drawing an isometric view on grid paper, again use the grid lines as guides for the correct axes of the view (Fig. 1.5). Remember to establish the position of each end of the line first when showing sloping lines.

Fig. 1.6

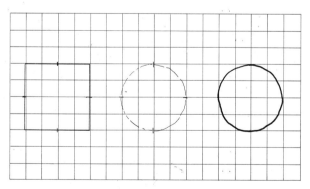

When drawing freehand circles and arcs, mark the position of points on the curve first and then join them as shown in Fig. 1.6. A paper trammel can be used to draw circles and arcs on plain paper (Fig. 1.7).

Fig. 1.7

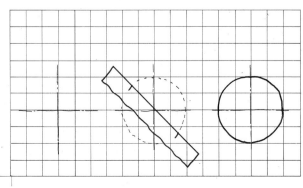

Freehand drawing

Fig. 1.8

Always sketch circles and arcs with your hand *inside* the curve as shown in Fig. 1.8.

Fig. 1.9

On plain paper, some vertical and horizontal lines can be drawn by holding the hand and pencil rigidly and sliding the finger tips along the edge of a board (Fig. 1.9).

Fig. 1.10

To draw irregular shapes in orthographic projection, such as the pipe clip shown in the photograph Fig. 1.10, use the boxing-in method illustrated in the sequence drawing (Figs 1.11, 1.12, 1.13 and 1.14).

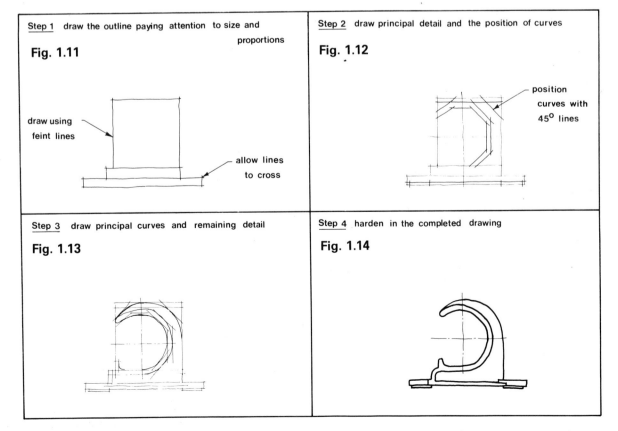

Step 1 draw the outline paying attention to size and proportions
Fig. 1.11
draw using feint lines
allow lines to cross

Step 2 draw principal detail and the position of curves
Fig. 1.12
position curves with 45° lines

Step 3 draw principal curves and remaining detail
Fig. 1.13

Step 4 harden in the completed drawing
Fig. 1.14

10 Freehand drawing techniques

When drawing inclined lines, particularly a pictorial view on plain paper, always keep your eye on the point to which you are drawing, as shown in the photograph.

Use the same boxing-in method for drawing pictorial views as shown in the sequence drawings of the flywheel (Figs 1.15, 1.16, 1.17 and 1.18).

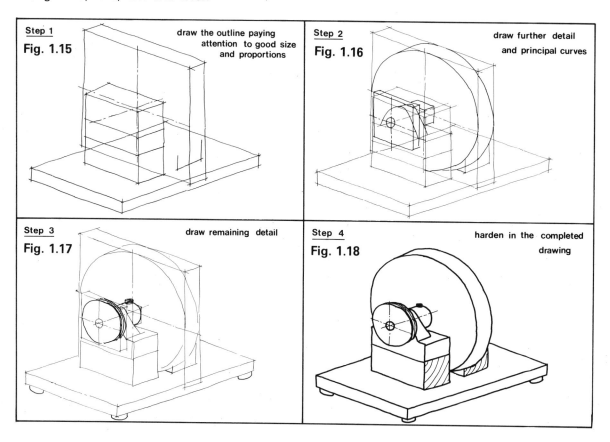

Step 1
Fig. 1.15   draw the outline paying attention to good size and proportions

Step 2
Fig. 1.16   draw further detail and principal curves

Step 3
Fig. 1.17   draw remaining detail

Step 4
Fig. 1.18   harden in the completed drawing

Freehand drawing techniques

A4 each

Fig. 1.19

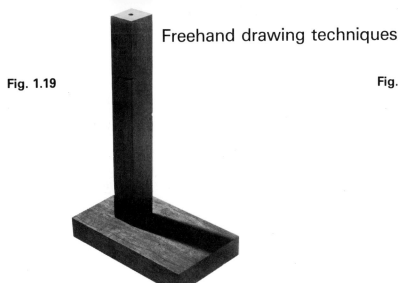

Fig. 1.20

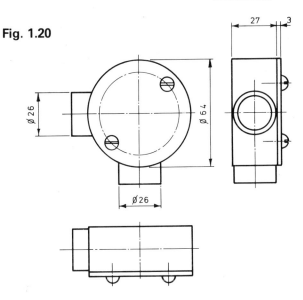

1. The photograph, Fig. 1.19, shows the basic construction of a wooden table lamp. The base of the lamp is 160 mm long, 100 mm wide and 25 mm thick, and the post is 30 mm square and 250 mm tall. There is a hole of 6 mm diameter through the centre of the post.

   Draw freehand orthographic views of the table lamp approximately half full size, and include in your drawings:

   (a) a suitable construction for joining the post to the base;
   (b) any shaping of the parts which you consider will improve the appearance of the lamp, remembering that it is made of wood;
   (c) a method for taking the electrical flex from the 6 mm hole in the post out to the back of the lamp.

2. Orthographic views of a junction box for electrical conduit are given in Fig. 1.20. Draw a freehand pictorial view of the fitting so that the maximum amount of detail is seen.

Fig. 1.22

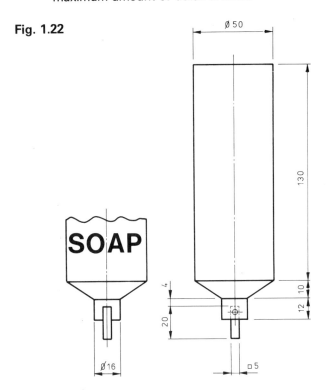

Fig. 1.21

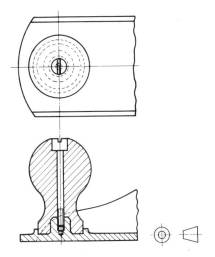

3. An illustration for a do-it-yourself magazine is to show the way in which the handle of a steel plane is attached to the body. Two views of the front of a steel plane, drawn in third angle orthographic projection, are given in Fig. 1.21.

   Draw freehand, approximately twice the size given, a pictorial view showing the parts exploded suitable for the illustration. You can use colour or shading to make your drawing more effective.

4. The part front view and completed side view of a plastic soap bottle as used in a restaurant cloakroom are given in Fig. 1.22. The bottle is to be mounted on a wall in such a way that soap is deposited on the hands when the spout at the base is pulled forward.

   Design and draw freehand a simple fitting to hold the soap bottle. Your design should show:

   (a) how the bottle is held securely in use;
   (b) a method of attaching the fitting to the wall;
   (c) a simple way of removing and replacing the bottle in the fitting.

   How has the fact that the bottle is made of soft plastic affected your design?

12 Freehand drawing techniques

5 Two parts of the handle of a drill are given in Fig. 1.23.

(a) Make freehand orthographic views of the assembled handle showing how you would attach the plastic knob to the metal strip so that the knob can revolve freely.

(b) Include a suitable shape for the knob. Remember that it is made of plastic and is to be held in the hand.

Fig. 1.23

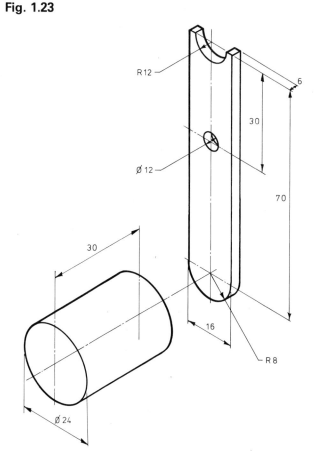

6 Orthographic views of a rainwater head are given in Fig. 1.24. Make a pictorial, scale 1:2, drawing of the fitting. Use isometric grid paper if you wish.

Fig. 1.24

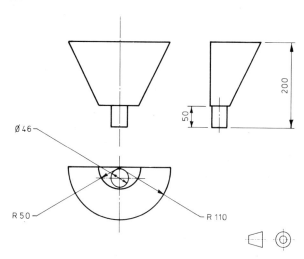

7 A wooden bracket which forms the basis of a wall-mounted light fitting is shown in Fig. 1.25.

(a) Make a freehand drawing of the bracket in orthographic projection, altering the shape in a way suitable for a wooden fitting.
(b) Complete the views by adding suitable fittings and a shade.
(c) How would you attach the fitting to the wall? Incorporate your suggestions in the drawing.

Fig. 1.25

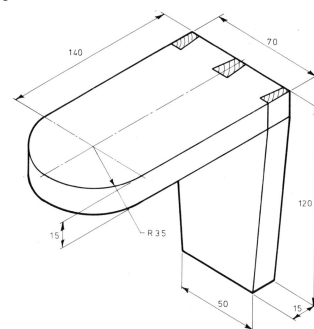

A3 each

2 Drafting methods

Sequence drawings

The technique of sequence drawings can be used to illustrate a process. A series of drawings, produced either freehand or using instruments, illustrates the steps required to complete the process.

Figure 2.1 is a series of sequence drawings. It illustrates a suggested sequence of steps for using the bracket shown in Fig. 2.2 to join two pieces of plastic-faced chip board.

In this type of drawing, it is essential that: the most effective type of projection is used; the best view is shown.

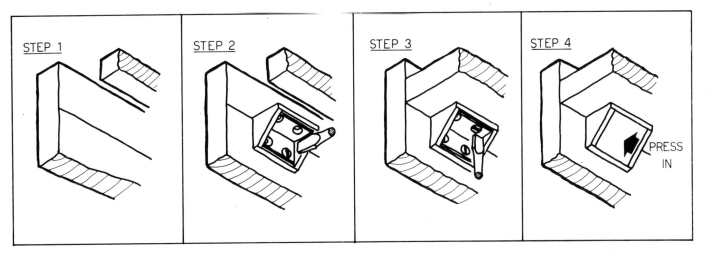

Fig. 2.1

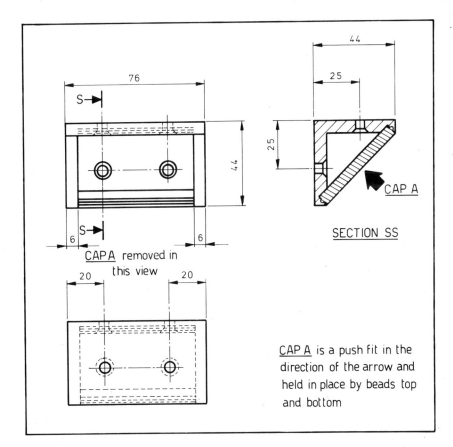

CAP A is a push fit in the direction of the arrow and held in place by beads top and bottom

Fig. 2.2

14 Sequence drawings – examples

A3 each

1. A do-it-yourself manual has a section on replacing the worn washer of a tap. One of the illustrations showing a full-size section of the tap is given in Fig. 2.3.
Draw, scale 2:1, with or without the use of instruments, a series of drawings to illustrate how to replace the worn washer. Before you start, it may be necessary for you to examine a tap to see how it works. Remember to turn off the water supply before you start!

Fig. 2.3   Section of a Pillar Tap

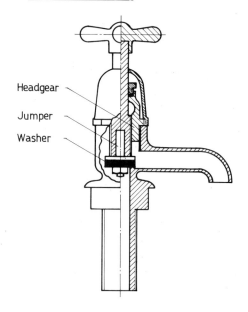

2. A breakfast cereal packet has a cut-out model of a windmill printed on it (Fig. 2.4). With the illustration is to be printed a series of five stages to help young children assemble the windmill. The five stages are as follows.

   (a) Cut out the pieces using scissors.
   (b) Fold the main body along the dotted lines. Glue tabs on the inside. Glue on the roof.
   (c) Fold the base along the dotted lines. Glue tab on the inside.
   (d) Glue base to main body.
   (e) Attach the sail with a paper fastener. Decorate your windmill as you wish.

   Draw five squares each 120 mm square, and use each square to illustrate one of the five stages.

Fig. 2.4

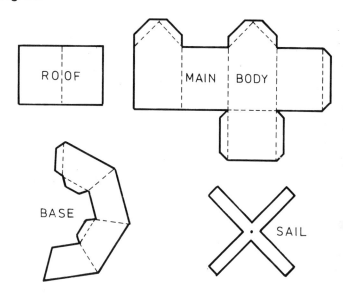

3. A sequence drawing showing how to replace the fuse of a 13 amp plug is required for a magazine in which it is assumed that the reader has little knowledge of electrical fittings.
Three photographs of the plug are given in Fig. 2.5. The cap is removed by undoing the retaining screw, and then the faulty fuse can be prised from its retaining clips and replaced with a new one.
Sketch freehand illustrations for replacing the fuse.

Fig. 2.5

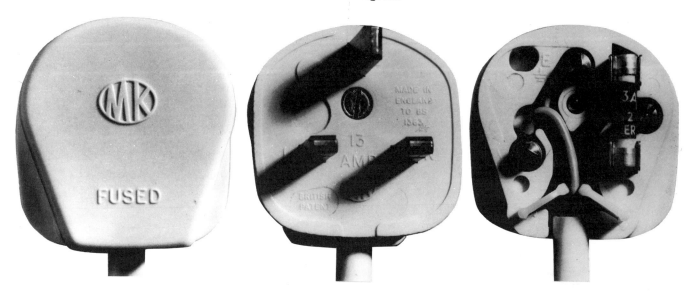

Drawing techniques — Exploded views

Fig. 2.6

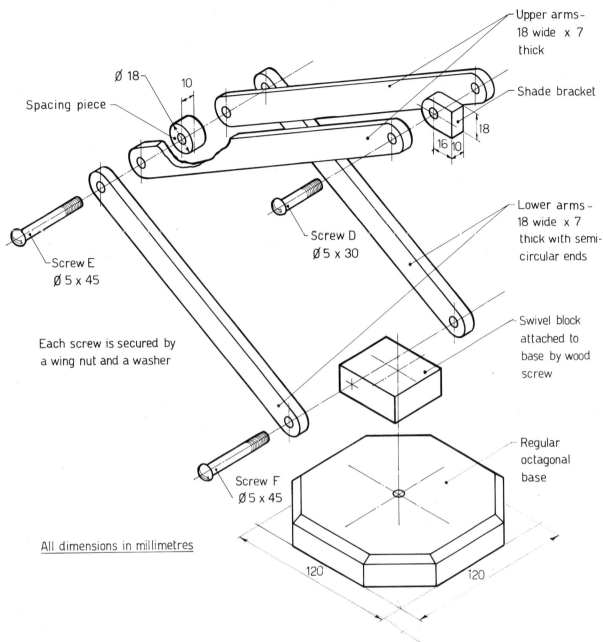

An example of an exploded pictorial view of a wooden, adjustable table lamp is given in Fig. 2.6.
Note
The parts are exploded or taken apart in the same direction as they would be taken apart in the actual model.
The use of written information to assist with the interpretation of the drawing.
The method of cutting-away parts in order to give a clearer picture of what otherwise would have been hidden.
The principle of exploded views can be applied equally to other pictorial forms of projection and to orthographic projection.

16 Exploded view — examples

Fig. 2.7

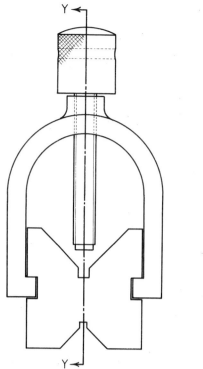
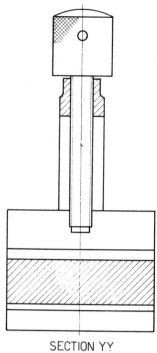

SECTION YY

1. Orthographic views of a 'V' block and clamp are given in Fig. 2.7. Draw an exploded isometric view of the 'V' block and clamp making each dimension twice the size printed.

2. Orthographic views of part of a wooden gate are given in Fig. 2.8. Draw an exploded pictorial view of the part of the gate making each dimension twice the size printed.
You will need to set up your A3 sheet of paper with the short edge horizontal for this drawing.

Fig. 2.8

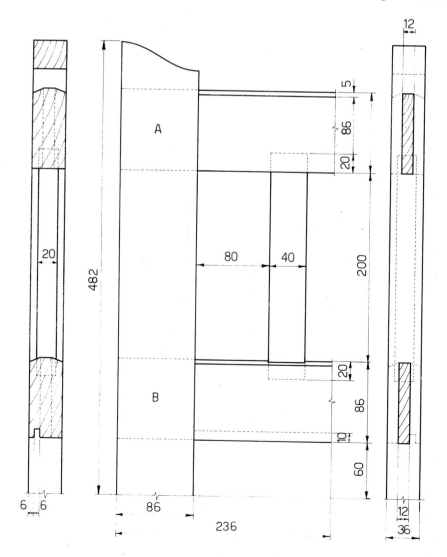

3 Applied geometry 1

Division of a line

To divide the line AB into five equal parts (Fig. 3.1)

Draw a line inclined to AB at approximately 20°.
Using compasses, mark five equal divisions, approximately one fifth the length of AB, along the inclined line.
Join the last point to B.
Draw lines parallel to this line through the other points.

Use a similar method for any number of equal divisions.

Fig. 3.1

Avoid constructions like those shown in Fig. 3.2.

Fig. 3.2

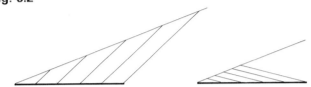

To divide the line CD in the ratio of 5:2 (Fig. 3.3)

Use the above construction to divide CD into 5 + 2 = 7 equal parts. Fig. 3.3. shows the completed construction.

Fig. 3.3

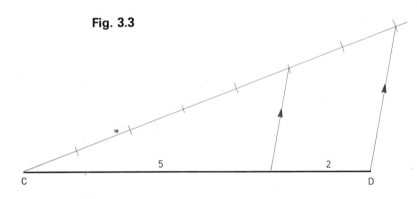

To construct the triangle EFG given the perimeter HI and the ratio of the sides 3:2:4 (Fig. 3.4)

Draw a line equal to the perimeter HI.
Divide the line into 3 + 2 + 4 equal parts (the sum of the ratio).
Use compasses to complete the triangle as shown in Fig. 3.4.

Fig. 3.4

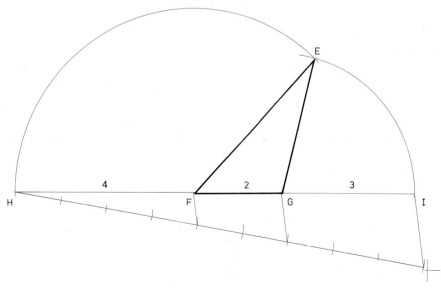

Construction of angles

To construct a line perpendicular to a given line at a given point J (Fig. 3.5)

Using compasses, draw arcs at an equal distance either side of J on the given line.
With these points as centres, draw arcs of convenient size to intersect as shown in Fig. 3.5.
Join the points of intersection to J to complete the required perpendicular line.

Fig. 3.5

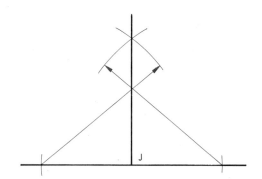

Where the given point is at the end of the given line, extend the line and proceed as in Fig. 3.6.

Fig. 3.6

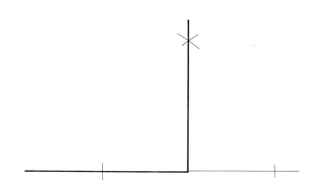

To construct a line perpendicular to a given line from a point K not on the line (Fig. 3.7)

Draw an arc, centre K, to intersect with the given line.
With centres on the points of intersection, draw two further arcs to intersect as shown.
Join the points of intersection to complete the required line.

Fig. 3.7

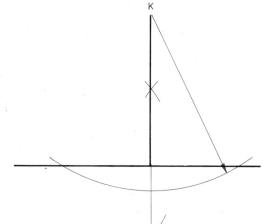

To construct an angle of 45° (Fig. 3.8)

Draw any line and mark the point L where the angle is to be constructed.
Draw a line perpendicular to the given line at L.
Bisect the 90° angle as shown in Fig. 3.8.

Fig. 3.8

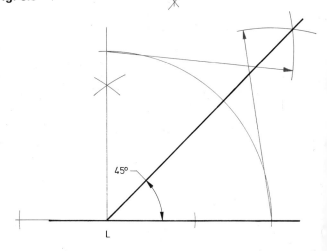

Construction of angles

Fig. 3.9

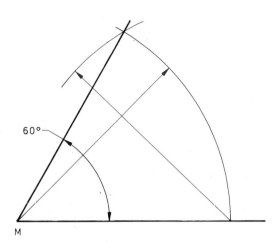

To construct an angle of 60° (Fig. 3.9)

Draw a line and mark the point *M* where the angle is to be constructed.

Draw an arc of convenient size with centre *M*.

Draw an arc of the same radius from the point of intersection of the first arc with the line.
Join *M* to the point of intersection of the two arcs to complete the required angle.

Fig. 3.10

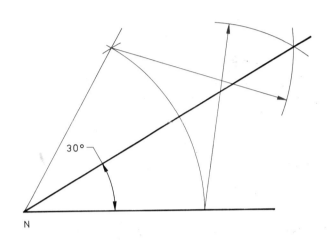

To construct an angle of 30° (Fig. 3.10)

Construct an angle of 60° and bisect it.

Fig. 3.11

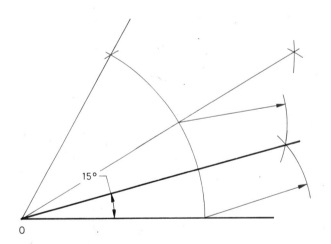

To construct an angle of 15° (Fig. 3.11)

Construct an angle of 30° and bisect it.

Fig. 3.12

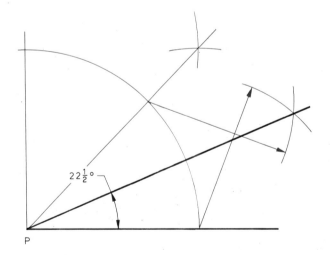

To construct an angle of 22½° (Fig. 3.12)

Construct an angle of 45° and bisect it.

How many other angles can you construct by combining the five constructions shown above?

Construction of regular polygons

Fig. 3.13

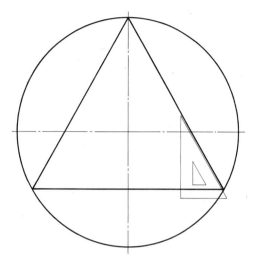

To inscribe an equilateral triangle in a circle (Fig. 3.13)

Starting at the top, use a 60° set square to draw the two sides of the triangle.
Complete the triangle with a horizontal line.

Fig. 3.14

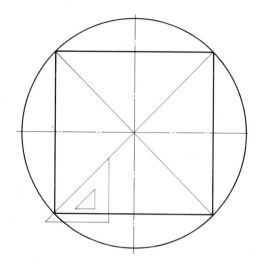

To inscribe a square in a circle (Fig. 3.14)

Draw 45° lines through the centre of the circle.
Complete the square with horizontal and vertical lines.

Fig. 3.15

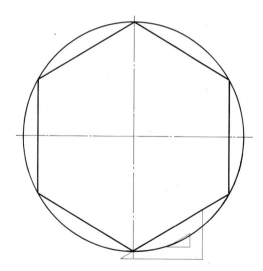

To inscribe a regular hexagon in a circle (Fig. 3.15)

Starting at the required point, use a 30°/60° set square to draw the sloping lines.
Complete the hexagon with horizontal or vertical lines.

Fig. 3.16

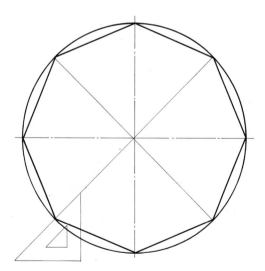

To inscribe a regular octagon in a circle (Fig. 3.16)

Draw 45° lines through the centre of the circle.
Complete the octagon as shown in Fig. 3.16.

Construction of regular polygons

Fig. 3.17

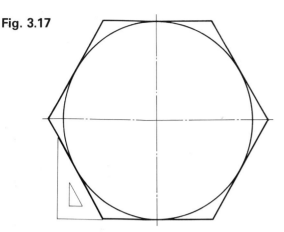

To construct a regular hexagon circumscribing a circle (Fig. 3.17)

Use a 30°/60° set square to draw the sloping sides tangential to the circle.
Complete the hexagon with horizontal or vertical lines.

Fig. 3.18

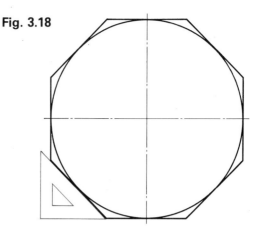

To construct a regular octagon circumscribing a circle (Fig. 3.18)

Use a 45° set square to draw the sloping lines tangential to the circle. Complete the octagon with horizontal and vertical lines.

Fig. 3.19

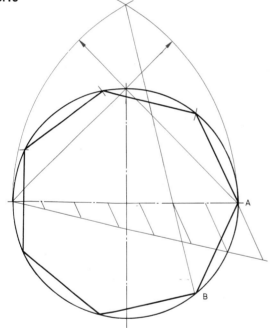

To construct any regular polygon in a circle (Fig. 3.19)

Divide the diameter of the circle into the same number of equal parts as the polygon is to have sides (seven in this example).

Draw two arcs to intersect with each other. The arcs have radii equal to the diameter of the circle; their centres are the ends of the diameter.

Join the point of intersection to the *second* division and extend the line to intersect with the circle at B.

Join AB to make one side of the required polygon.

Complete the polygon using compasses.

Note: Always use the second division, irrespective of the number of sides required.

Fig. 3.20

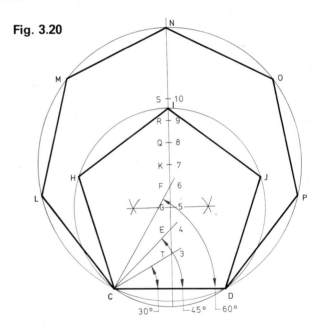

To construct any regular polygon on a given line (Fig. 3.20)

Draw the line CD.

Draw the perpendicular bisector of the line.

Draw a line through C at 45° to CD to intersect with the perpendicular at E.

Draw a line through C at 60° to CD to intersect with the perpendicular at F.

Bisect EF at G.

Draw a circle centre G to pass through C and D. Using compasses, make CH = HI = IJ = JD = CD to form a *regular pentagon*.

Make KF = FG

Draw a circle centre K to pass through C and D.

Using compasses, make CL = LM = MN = NO = OP = PD = CD to form a *regular heptagon*.

Examples based on lines, angles and polygons

Fig. 3.21   **Fig. 3.22**   **Fig. 3.23**

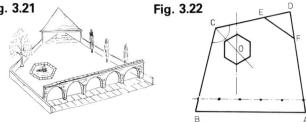
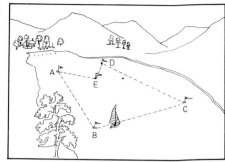

1. A small plot of land is laid out as a public garden as shown in the sketch (Fig. 3.21) and the outline plan (Fig. 3.22).
   Draw, to scale 1:200, a plan of the outline of the garden given the following dimensions in metres.

   AB = 33·6   BC = 25·2   ED = 7·6
   AC = 36·6   DA = 29·8   CD = 22·8
   EF = 10·8

   The arches are equi-spaced, parallel to the front edge of the plot AB and 4 m back from it. The pond is a regular hexagon with sides of 5 m. The centre of the pond is 10 m from the corner C and lies on the bisector of the angle at C.

2. A yacht club situated at the end of a lake is shown in the sketch (Fig. 3.23). Five buoys are used to mark a short training course for dinghies. To scale 1:1000, make a drawing of the course given:
   AB = 205 m   AE = 90 m   AD = 95 m
   CD = 143 m   BC = 138 m
   The angles BAE = 45° and ABC = 60° must be constructed.

Fig. 3.24

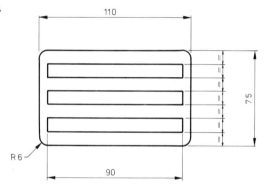

Fig. 3.25

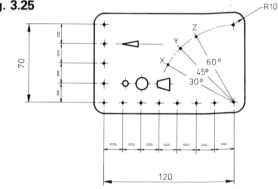

3. A view of a ventilator plate for a gas refrigerator is given in Fig. 3.24. Draw the given view full size.

4. A plastic drawing template is shown in Fig. 3.25.
   (a) Draw the given outline, scale 1:1.
   (b) Use a geometrical construction to find the centre of each of the holes on two of its sides.
   (c) Find, by construction, the position of the three holes X, Y and Z.

Fig. 3.26

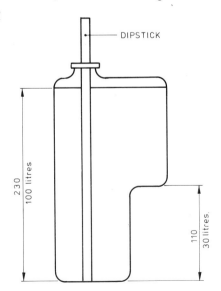

Fig. 3.27

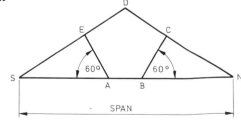

5. A cross-section of a fuel tank complete with dipstick and the depth and quantity of fuel at two levels is given in Fig. 3.26. Draw, scale 1:1, a suitable dipstick to show the amount of fuel in the tank at 10 litre intervals. The dipstick is made of 10 mm diameter rod.

6. The roof structure of a small factory is triangular in section (Fig. 3.27). Within the triangle are two supports which form part of the irregular polygon ABCDE.
   The sides of this polygon are in the ratio:
   AB : BC : CD : DE : EA as 2 : 3 : 3 : 3 : 3
   The perimeter of the polygon is 12 m.
   Find, by drawing to scale 1:50, the actual span of the roof.

Examples based on lines, angles and polygons

Fig. 3.28

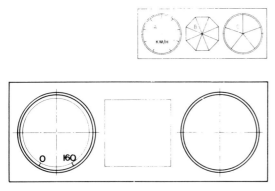

7 A view of the instrument panel of a car is given in Fig. 3.28.
Copy the larger view twice the size of the illustration and complete the drawing given:

(a) the speedometer at *A* is graduated from 0 – 160 km/h in equal 20 km/h units, add numbers to the graduations;
(b) the warning light cluster *B* is a regular octagon fitting exactly in the square;
(c) the spaces for the five gauges in *C* are of equal size.

Fig. 3.29

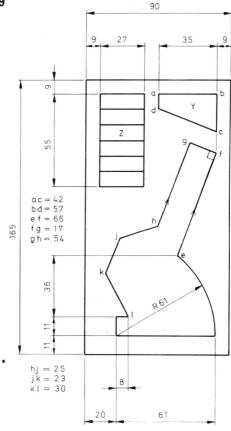

8 A container for a microscope set is given in Fig. 3.29. Draw the outline of the container to the dimensions given. The divisions in *Z* are of equal size. Draw the largest circle to fit in recess *Y*.

Fig. 3.30

Fig. 3.31

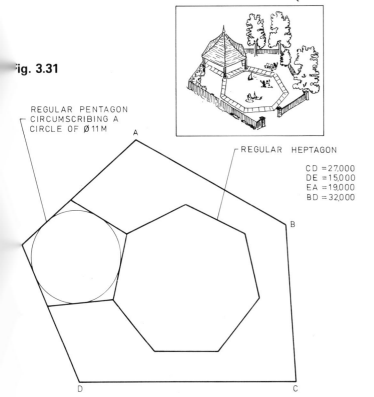

REGULAR PENTAGON CIRCUMSCRIBING A CIRCLE OF Ø 11 M

REGULAR HEPTAGON
CD = 27.000
DE = 15.000
EA = 19.000
BD = 32.000

9 A children's paddling pool and adjacent shelter are to be built on an irregularly shaped site. The sketch, Fig. 3.30, gives an impression of the site, and a dimensioned plan view of it is given in Fig. 3.31. Use geometrical constructions to draw a plan view of the site, pool and shelter to scale 1:200.

Fig. 3.32

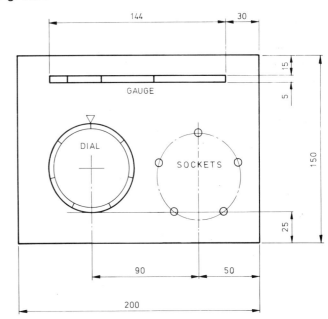

10 The face of a control box is shown in Fig. 3.32. The strip *gauge* is graduated in the ratio of 1:2:3:4. The periphery of the *dial*, diameter 70 mm, is divided into seven equal parts as shown. The five multi-point *sockets* are 40 mm apart and cyclic. Use geometrical constructions to draw the control box.

4 Electrical circuits

Fig. 4.2

Conductor

Junction of conduction

Switch

Diode

Push button switch (break contact)

Push button switch (make contact)

Fuse

Plug & socket

Ammeter

Voltmeter

Filament lamp

Fixed resistor

Variable resistor

Electrical bell

Electrical buzzer

Earth

Battery of cells

Transformer

Capacitor

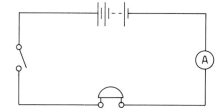
Transistor

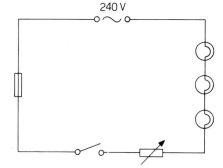

Fig. 4.1 The photograph, Fig. 4.1, shows a battery-operated bell system in which an ammeter has been included. When the switch is turned on, the bell rings and a reading is indicated on the ammeter.

In order to save time in drawing such a circuit, each component part is represented diagrammatically. British Standard symbols are used to represent each component (Fig. 4.2). The complete list is available in BS 3939 (Graphical symbols for electrical power, telecommunications and electronics diagrams) and may be obtained from the British Standards Institution, 2, Park Street, London, W1A 2BS. An abbreviated list is available in PD 7307:1982.

Fig. 4.3

A graphical method of representing the electrical circuit shown in the photograph is given in Fig. 4.3. It is usual for the circuit to be in rectangular form with the conductors drawn vertically or horizontally wherever possible.

Fig. 4.4

Here is a second example of an electrical circuit of lighting for a school hall in which three spotlights, powered from the mains supply, can be dimmed using a variable resistor. The circuit also contains a fuse and a switch (Fig. 4.4). Note the symbol for a 240 V mains supply.

Electrical circuits – examples

A4 each   25

Fig. 4.5

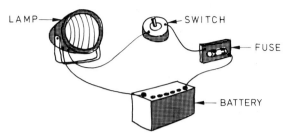

Fig. 4.6

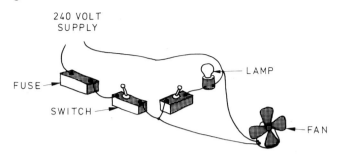

1. The sketch of a circuit for a spotlight of a motor boat is given in Fig. 4.5.
   Using conventional symbols, draw a diagram of the circuit.

2. A bell system consists of an electric bell operated by a push switch at the front door of the house, and a buzzer, operated by a push switch, at the back door. Both systems use a 12-volt supply from a transformer which obtains its power from the mains supply of 240 volts. A fuse is fitted between the mains supply and the independently earthed transformer.
   Design and draw a suitable electrical circuit.

3. A stage spotlight circuit consists of a spotlight, a fuse and a dimmer (variable resistor) operated from a mains supply of 240 volts.
   Draw the circuit.

4. A light inside a wardrobe comes on when the door of the wardrobe is opened. The light is powered through a transformer from a 240-volt mains supply. There is a switch and a fuse between the mains supply and the transformer.
   Use appropriate symbols to draw the circuit.

5. A sketch showing the electrical circuit of an overhead projector is given in Fig. 4.6. The fan is earthed to the supply through the case of the projector.
   Draw a wiring diagram of the circuit.

Fig. 4.7

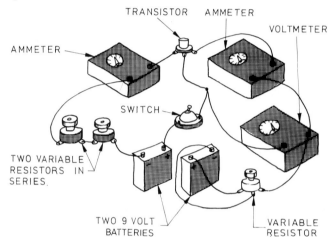

6. Fig. 4.7 shows the assembled components for a transistor testing device.
   Draw an electrical circuit for the device.

HONDA CB400A IGNITION AND STARTING SYSTEM

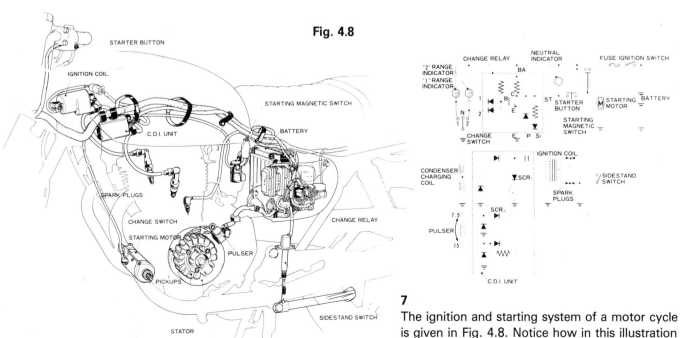

Fig. 4.8

7. The ignition and starting system of a motor cycle is given in Fig. 4.8. Notice how in this illustration both symbols and text are used to explain the pictorial view.

5 Applied geometry 2

Enlargement and reduction

Fig. 5.1

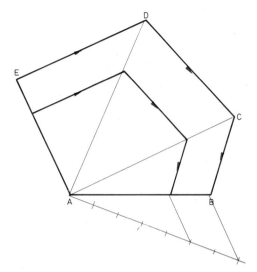

First method

To reduce a plane figure ABCDE to a similar figure having each side five-sevenths the length of the corresponding sides of ABCDE (Fig. 5.1).

Divide any side (AB) in the ratio of 5:2 (Fig. 3.3).

Join A to D and C.

Draw lines parallel to BC, CD and DE to complete the required figure.

Fig. 5.2

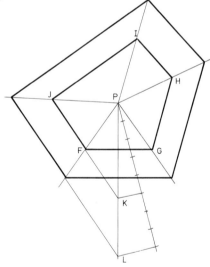

Second method

To enlarge a plane figure FGHIJ to a similar figure having each side larger in the ratio of 8:5 to the corresponding sides of FGHIJ (Fig. 5.2). Select any point P inside the polygon.

Join P to the corners of the polygon and extend outwards.

Draw a vertical line from P to meet JF extended at K.

Make KL three-fifths PL, i.e. so that PL:PK = 8:5.

Complete the figure as shown.

Fig. 5.3

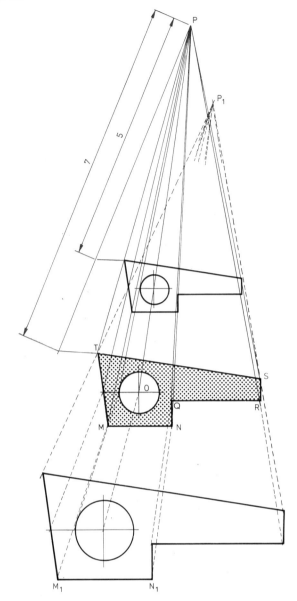

Third method

This is a method similar to 2 except that P is at any convenient distance away from the given figure MNQRST (Fig. 5.3).
Note the construction for finding the centre of the reduced circle.

To draw an enlarged figure given the new length of MN (M1N1). Join M1 to M and N1 to N and extend the lines to meet at P1 and then proceed as before.

Enlargement and reduction – examples

A3 each

Fig. 5.4

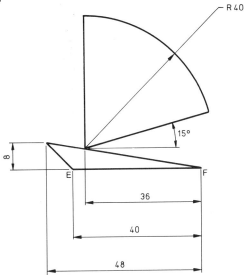

1. You are required to make a trophy for a sailing club.
   The central motif of the trophy is a flat silver plate in the shape of a regular pentagon. Inside the pentagon is an enamelled crest which is in the shape of the club's lapel badge, only larger (Fig. 5.4).
   Draw, scale 1:1, a pentagon with each side 90 mm, and in it draw an enlarged view of the badge with the line *EF* 70 mm long and 20 mm above the horizontal base of the pentagon.

Fig. 5.7

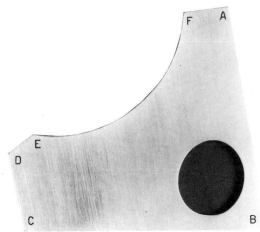

3. The photograph Fig. 5.7, shows a mounting plate for a television set. Given: *AB* = 94 mm; *BC* = 102 mm; *CD* = 27 mm; *DF* = 86 mm; angle *BCD* = 105°; angle *FAB* = 90°; *FB* = 96 mm; the arc *EF*, radius 50 mm, is tangential to the line joining *A* to *C*; the hole, diameter 30 mm, has its centre on the bisector of the angle *ABC* and is 35 mm from *B*; *E* lies on the straight line *DF*.

   (a) Draw the plate scale 1:1 with *BC* horizontal.
   (b) Draw a similar figure with the side corresponding to BC = 62 mm.

Fig. 5.5

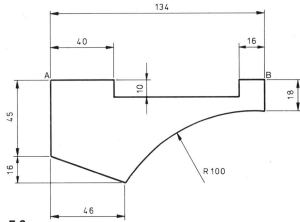

Fig. 5.6

2. The engine cowling of a model Spitfire kit has been lost as seen in the photograph Fig. 5.5, and has to be made from a sheet of thin plastic. The only available drawing of the part is too large and has to be reduced in size.
   The available drawing is given in Fig. 5.6. Copy the given view and from it make a reduced drawing of the part with *AB* 80 mm long.

Fig. 5.8

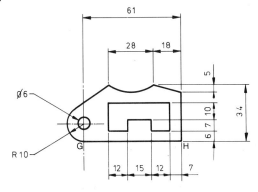

4. The outline of a tumbler of a lock is given in Fig. 5.8. Draw this figure and then draw a similar figure with *GH* 43 mm.

6 Charts

A flow chart represents graphically the series of stages or operations in a process. There is a recommended procedure for doing this and it is specified in BS 4058 and PD 7307, from which the following symbols have been extracted. Further information should be obtained from the complete British Standard.

An abbreviated description or heading of each stage in the process is printed in the appropriate symbols which are in turn joined by flow lines in the order in which they take place.

Some of the principal symbols for data processing flow charts are shown in Fig. 6.1; a simple example of a flow chart is given in Fig. 6.2.

It should be noted that often there is more than one way of designing and constructing a flow chart for the same process.

Fig. 6.1

Process
This symbol represents any kind of processing function.

Preparation
This symbol represents modification or preparation.

Terminal/interrupt
This symbol represents a terminal point in a flow chart, i.e. a stop, halt, delay or interrupt.

Input/output
This symbol represents an addition or subtraction in the process.

Decision
This symbol represents a decision involving a number of alternative paths to be followed.

Flow charts

Flow lines
These lines should not cross. Joining flow lines should be staggered.

Example

Stages in checking a non-functioning 13 amp electrical plug in which it is believed that the fuse has blown.

(a) Remove the plug from its socket.
(b) Unscrew the retaining bolt and remove the back.
(c) Check that the electrical cables are connected to their correct terminals.
(d) If they are not connected, re-connect as required.
(e) Replace the back of the plug and tighten the retaining bolt.
(f) Test to see if the plug now functions correctly.
(g) If it does, then the process is finished.
(h) If it does not, unscrew the retaining bolt and remove the back.
(i) Take out and replace the fuse.
(j) Replace the back and tighten the retaining bolt.
(k) Test to see if the plug now functions correctly.
(l) If it does, then the process is finished.
(m) If it does not, then decide on a further course of action.

Fig. 6.2

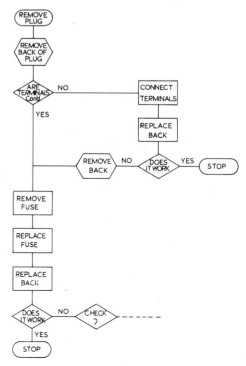

Flow charts – examples

A4 each

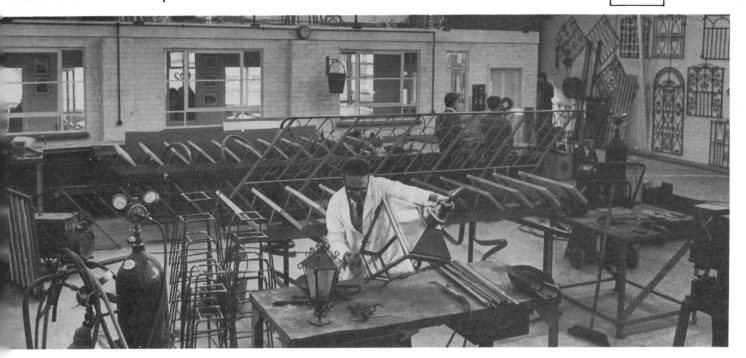

1. A company making decorative lanterns requires a flow chart to be used for testing their products.

    The operations involved are:

    (a) fit a 13 amp plug to the cable of the lantern and plug in to the mains supply;
    (b) fit a light bulb in the bulb-holder of the lantern;
    (c) switch on the mains supply;
    (d) if the bulb lights up –
    (e) turn off the mains supply;
    (f) remove the plug from the socket;
    (g) remove the plug from the cable – test completed;
    (h) if the bulb does not light up –
    (i) change the bulb;
    (j) if the bulb still does not light up;
    (k) turn off the mains supply;
    (l) remove the plug from the socket.
    (m) check that the plug has been correctly wired;
    (n) if it has not, connect the wires correctly and return to (b);
    (o) if the plug has been correctly wired –
    (p) check that the wiring to the bulb holder is correct;
    (q) if it is not, connect the wires correctly and return to (b).

    Using standard symbols, draw a suitable flow chart.

2. The tyre of a bicycle is found to be flat. The stages for repairing the tyre are given below. Using standard symbols, draw a flow chart showing how the tyre would be repaired.

    (a) Rear tyre flat.
    (b) Unscrew valve cap.
    (c) Check valve.
    (d) Replace valve.
    (e) Pump up tyre.
    (f) Tyre remains inflated; finish.
    (g) Tyre goes down.
    (h) Remove valve.
    (i) Remove valve ring.
    (j) Remove outer cover.
    (k) Remove inner tube.
    (l) Find the puncture.
    (m) Repair puncture.
    (n) Apply french chalk.
    (o) Replace inner tube.
    (p) Replace outer cover.
    (q) Replace valve ring.
    (r) Replace valve.
    (s) Pump up tyre.
    (t) Tyre remains inflated; finish.

3. To prepare an automatic washing machine for use, the following procedure should be followed. Unscrew the screws holding the rear panel of the machine in place; take out the packing piece marked *A*. Unscrew the screws holding the lid in place, take off the lid; take out the packing piece marked *B*. Make quite certain both packing pieces have been removed. Replace both the rear panel and the lid.

    Now connect the ends of the water filling hoses to the machine. The red fitting must be screwed to the red threaded outlet at the rear of the machine; the other fitting must be screwed to the white threaded outlet at the rear of the machine. The ends of the two filler hoses should now be connected to the water supply. Red to hot water, white to cold water. Turn on the two water supply taps.

    Make sure the electric plug fitted to the electric cord is a 13 amp plug. Ensure that the fuse is for 13 amps. Plug into a 13 amp wall socket. Switch the machine on. The machine is now ready to operate.

    Using standard symbols, draw a flow chart describing the above procedure.

Time charts

Fig. 6.3

DISS HIGH SCHOOL		1982																				
GREENGRASS BUILDERS		MAY				JUNE				JULY				AUG.				SEPT.				
CLIENT	WEEK COMMENCING	3	10	17	24	31	7	14	21	28	5	12	19	26	2	9	16	23	30	6	13	20	27
ARCHITECT	NUMBER	1	2	3	4	5	6	7	8	9	10	11	12	13	14	15	16	17	18	19	20	21	22
1	FENCING, SITE SHEDS													H	H								
2	SETTING OUT																						
3	EXCAVATE TOP SOIL, FOUNDATIONS, DUCTS																						
4	CONCRETE FOUNDATIONS													O	O								
5	BRICKWORK TO D.P.C.																						
6	BRICKWORK OR CONCRETE TO DUCTS																						
7	HARDCORE SITE SLAB													L	L								
8	CONCRETE SITE SLAB																						
9	STEELWORK BRICKWORK TO ROOF LEVEL																						
10	CARPENTRY WOODWOOL DECKING													I	I								
11	FELT ROOFING R.W.P.																						
12	METAL WINDOWS																						
13	HEATING													D	D								
14	ELECTRICAL																						
15	DUCT COVERS																						
16	FLOORSCREEDS & CERAMIC TILES													A	A								
17	JOINER																						
18	PLUMBING																						
19	DECORATING													Y	Y								
20	FLOORFINISHES																						
21	HAND OVER INSPECTION																						
22	DRAINAGE						TOILET BLOCK																
23	ALTERATIONS TO EXISTING WOODWORK																						
24	WIDENING EXISTING ROUNDABOUT					MATERIAL WORKSHOP																	

Another method to illustrate graphically a sequence of operations of a manufacturing process or programme is the time chart (Fig. 6.3). A time chart includes not only the sequence of operations, but also when each part occurs and how long it takes.

This type of chart is best illustrated on a grid of horizontal and vertical lines on which the time allocated to each operation is represented by rectangular blocks. Note that it is possible to represent two or more separate sequences on the same chart; the sequences can be differentiated by the use of colour or shading.

Time chart – examples

Operation	From	To
Machining timber to size	March 1st (Monday)	March 3rd
Mortising legs	March 4th	March 9th
Tenoning rails	March 11th	March 12th
Edge-joining tops	March 15th	March 17th
Surfacing tops	April 1st	April 5th
Machining tops to size and shape	April 6th	April 7th
Moulding rails	April 18th	April 24th
Shaping legs	April 25th	April 30th
Dovetailing drawer sides	May 1st	May 7th
Grooving drawer sides and fronts	May 8th	
Sanding legs	May 2nd	May 3rd
Sanding rails	May 5th	May 6th
Gluing side frames	May 7th	May 21st
Gluing drawers	May 21st	May 28th
Fitting drawer bottoms	May 29th	
Gluing complete frame	May 29th	June 6th
Sanding complete frame	June 7th	June 11th
Fitting drawer	June 12th	June 14th
Fitting top	June 15th	June 16th
Polishing	June 19th	June 23rd
Packing	June 27th	June 30th

1 The programme for manufacturing a batch of small tables with drawers is given above.

Draw a suitable grid and on it construct a time chart illustrating the programme. Note that some of the processes take place at the same time.

2 A cricket pavilion is to be built on a village green. The work is to be sub-contracted to specialist tradesmen as listed below. The work to be undertaken and the time required to complete the work is also given. Given that the total time required to finish the building is 7 weeks and that there are 6 working days in each week, complete a time chart for the building programme, the outline of which is given below.

Note: Work must proceed in order, i.e. the roof cannot be put on until the walls have been completed. Other than obvious factors, the following conditions apply:

(a) the electrician fixes the cables in position before the walls are plastered, he returns later to put on switches, etc.
(b) the plumber cannot work until the plasterer has finished.
(c) the floor is concreted after the bricklayer has completed the wall up to that level.

Work and time for each tradesman

Labourer	digging foundations 2 days; concreting floors 2 days; assisting bricklayer while he is at work 14 days
Bricklayer	foundations and brickwork up to floor level 4 days; remaining brickwork including building in door and window frames 10 days
Roofing contractors	putting on roof 6 days
Electrician	wiring 1 day; fittings 1 day
Glazier	glazing windows 1 day
Plasterer	plastering ceiling and walls 5 days
Plumber	fitting pipework, sink and water heater 4 days
Joiner	fitting door, window, skirting board and other woodwork 4 days
Decorator	decorating throughout 4 days

A3 each

7 Applied geometry 3 — Circles

You need to know the names of the following parts of a circle, reading across pages 32 and 33.

Fig. 7.1

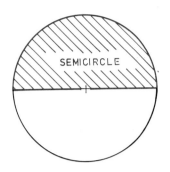

A *semicircle* is half of a circle (Fig. 7.1).

Fig. 7.2

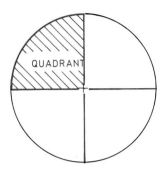

A *quadrant* is a quarter of a circle (Fig. 7.2).

Fig. 7.5

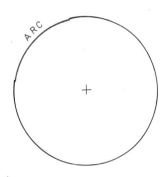

An *arc* is part of the circumference (Fig. 7.5).

Fig. 7.6

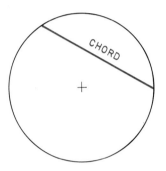

A *chord* is a straight line joining the extremities of the arc of a circle (Fig. 7.6).

Fig. 7.9

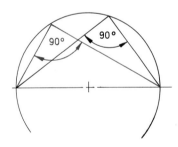

The angle subtended by the diameter of a circle is always a right angle (Fig. 7.9).

Fig. 7.10

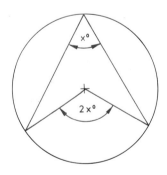

The angle subtended by an arc at the centre of a circle is always twice the size of the angle subtended by the same arc on the circumference of the circle (Fig. 7.10).

Fig. 7.13

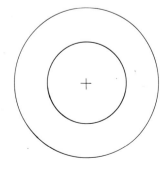

Concentric circles (Fig. 7.13)
Circles with a common centre are concentric.

Fig. 7.14

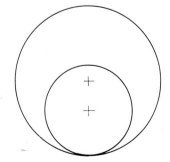

Eccentric circles (Fig. 7.14)
Circles without a common centre are eccentric.

Circles

Fig. 7.3

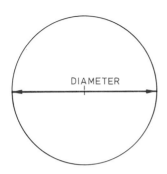

The *diameter* of a circle is a straight line passing through the centre of the circle from one side of it to the other (Fig. 7.3).

Fig. 7.4

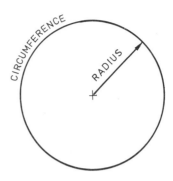

The *circumference* of a circle is the encompassing boundary of the circle (Fig. 7.4).

Fig. 7.7

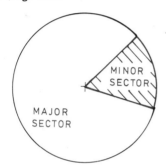

A *sector* is that part of a circle bounded by two radii and an arc (Fig. 7.7).

Fig. 7.8

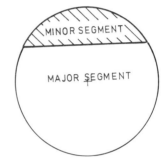

A *segment* is that part of a circle bounded by a chord and an arc (Fig. 7.8).

Fig. 7.11

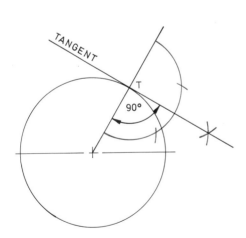

A straight line which touches the circumference of a circle at a point is called a *tangent*. The point at which the line touches the circle is called the *point of tangency*. (T).
A straight line from the centre of the circle to the point of tangency is called a *normal*. The angle between the tangent and the normal is 90° (Fig. 7.11).

Fig. 7.12

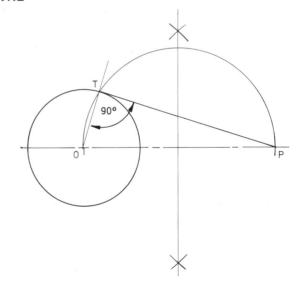

Tangent to a circle from a point P (Fig. 7.12)
Construction:

Bisect *OP*.

Draw the semicircle *OP*.

T is the point of tangency. Draw *TP*.

34 Circles – examples

1. A trade symbol used by Reckitt & Colman is shown in Fig. 7.15. Draw the symbol, scale 1:1. Start by drawing the square centre of the symbol, and subdivide one side of it into seven equal parts (see page 17) to obtain the radius of each part-circle.

Fig. 7.15

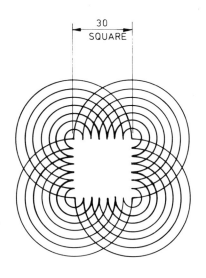

2. The logo for Goblin Ltd. consists of concentric circles (Fig. 7.16). Draw the logo, scale 1:1. Complete the logo as shown in the small illustration by adding the rectangle and the lettering.

Fig. 7.16

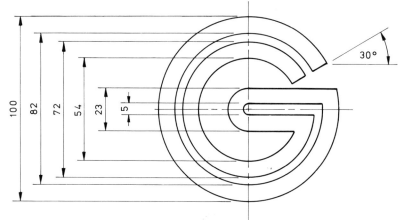

3. Part of the mechanism of an animated window display is shown in the sketch Fig. 7.17. Draw the given diagram Fig. 7.18, and then determine the maximum vertical movement of the end E of the rod for one revolution of the cam.

Fig. 7.17

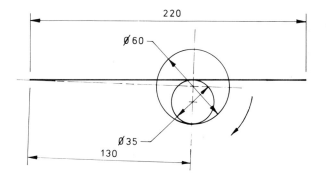

Fig. 7.18

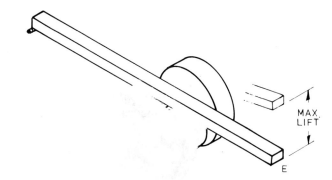

Tangents

Fig. 7.19

Fig. 7.20

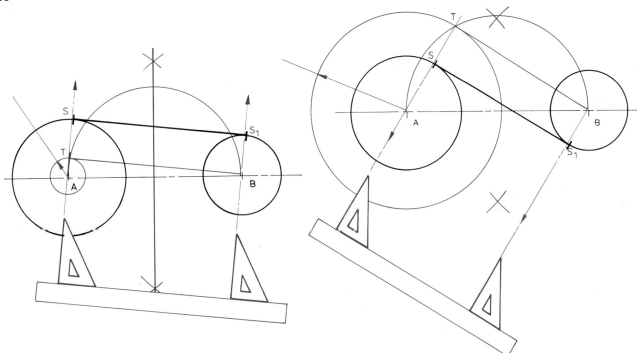

Exterior tangent between two circles of unequal diameter (Fig. 7.19)

Construction:

Refer to Fig. 7.12, page 33.

Draw circle centre *A* radius as shown.

Construct the tangent from this circle to *B*.

Mark the point of tangency *T*.

Draw the normal *AT*, and extend to *S*.

Draw *BS1* parallel to *AS*.

SS1 is the required tangent.

Interior tangent between two circles of unequal diameter (Fig. 7.20)

Construction:

Draw circle centre *A*, radius as shown.

Construct the tangent from this circle to *B*.

Mark the point of tangency *T*.

Draw the normal *AT*, and mark the point *S*.

Draw *BS1* parallel to *AS*.

SS1 is the required tangent.

Fig. 7.21

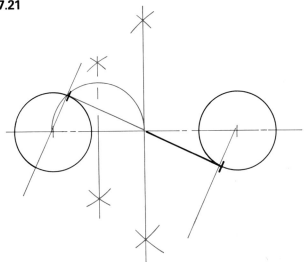

Fig. 7.22

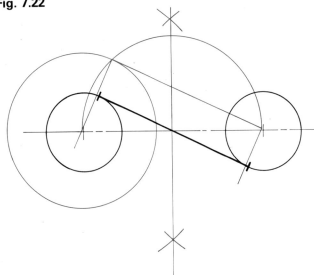

Two methods for drawing an interior tangent between two circles of equal diameter (Figs 7.21 and 7.22)

The constructions are self-explanatory.

36 Tangents – examples

Fig. 7.23

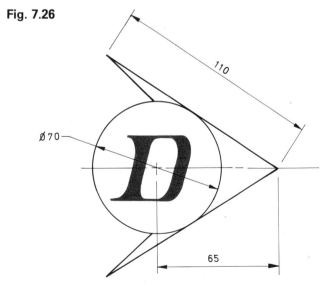

Fig. 7.24

Examples of tangents to circles are common. The photographs Figs 7.23 and 7.24 show a road sign and a logo which include tangents. Look about you for further examples, sketch them and then draw them to scale using the construction given on page 33.

Fig. 7.25

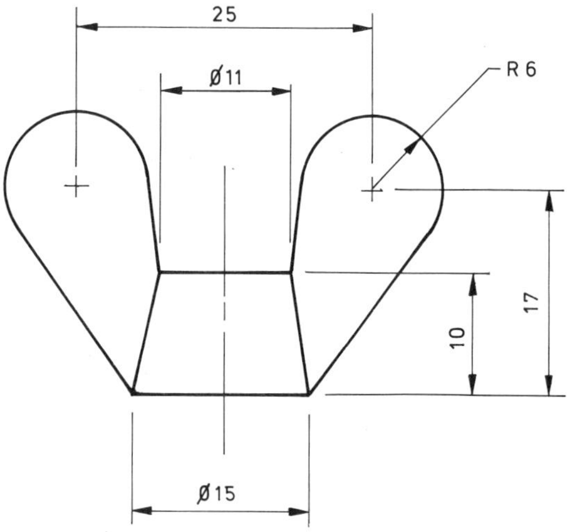

Fig. 7.26

Two further examples are given above.

1. The profile of a wing nut is shown in Fig. 7.25. Draw the given view scale 4:1 (four times the given size) using a geometrical construction for the tangents.

2. The Dunlop logo as seen on their sports equipment is shown in Fig. 7.26. Draw the logo, scale 1:1, using geometrical constructions for the tangents.

Tangential arcs

Fig. 7.27

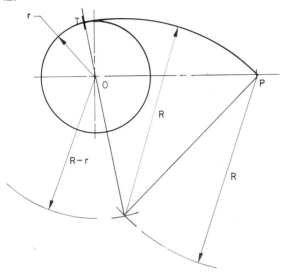

Inward arc from a point P to a circle centre O (Fig. 7.27)

Construction:

Draw an arc centre *P*, radius equal to that of the required arc *R*.

Draw an arc radius $R - r$ (radius of the given circle) so that the arcs intersect.

This is the centre of the required arc.
A line from this point to the centre of the given circle and extended locates the point of tangency *T*.

Fig. 7.28

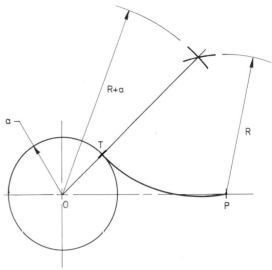

Outward arc from a point P to a circle centre O (Fig. 7.28)

Construction:

Draw an arc centre *P*, radius equal to that of the required arc *R*.

Draw an arc radius $R + a$ (radius of the given circle) so that the arcs intersect.

This is the centre of the required arc.
A line from this point to the centre of the given circle locates the point of tangency *T*.

Fig. 7.29

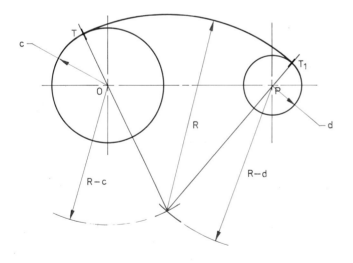

Inward common tangent to two unequal circles (Fig. 7.29)

Construction:

Draw an arc centre *O* radius $R - c$

Draw an arc centre *P* radius $R - d$

The point of intersection of the two arcs is the centre of the required arc.
T and *T1* are the points of tangency.

Fig. 7.30

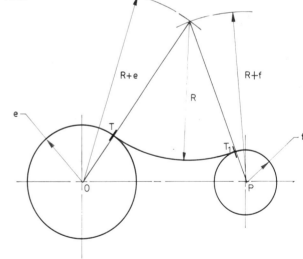

Outward common tangent to two unequal circles (Fig. 7.30)

Construction:

Draw an arc centre *O* radius $R + e$

Draw an arc centre *P* radius $R + f$

The point of intersection of the two arcs is the centre of the required arc.
T and *T1* are the points of tangency.

38 Tangential arcs – examples

A4 each

1. A view of a lever is given in Fig. 7.31.

   Draw a scale 1:1 view of the lever using geometrical constructions throughout.

Fig. 7.31

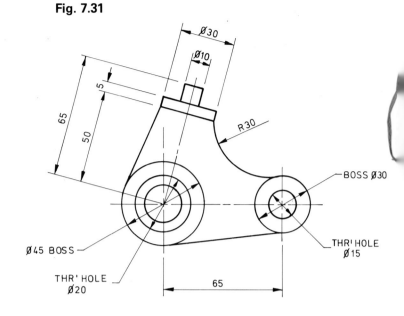

2. The outline of a quick release emergency handle is given in Fig. 7.32.

   Draw the outline, scale 1:1, using geometrical constructions for the tangential arcs.

   What does the shaded area on the handle signify?

Fig. 7.32

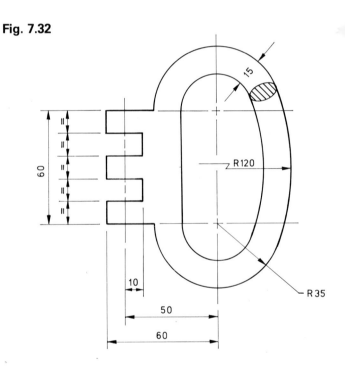

3. One of the four blades of an outboard motor propeller is shown in Fig. 7.33.

   Draw, scale 1:1, the outline of the blade and boss using a geometrical construction for each tangential arc.

Fig. 7.33

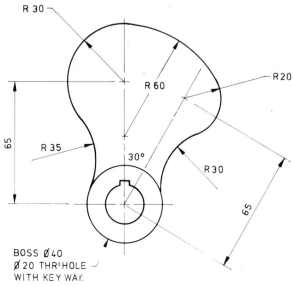

Tangential arcs – examples

A4 each

4. A partly-drawn view of a cylinder head gasket for a small engine is given in Fig. 7.34, together with a pictorial view of the gasket. Draw a scale 1:1 view of the gasket using geometrical constructions for the tangential arcs.

Fig. 7.34

Fig. 7.35

Fig. 7.36

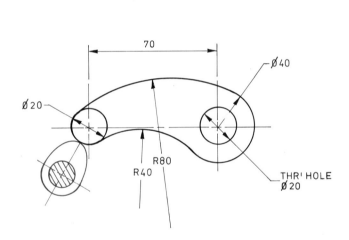

5. The outline of a cam follower is given in Fig. 7.35. The cam, shown in red, is mounted on a shaft and rotates clockwise. The follower is in direct contact with the cam and transmits a predetermined movement to the mechanism. Draw, scale 1:1, the outline of the cam follower using geometrical constructions for the tangential arcs.

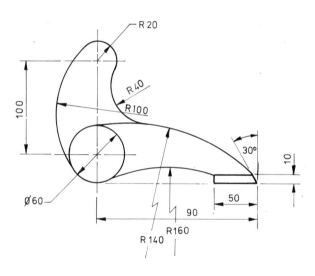

6. The outline of a ratchet pawl is given in Fig. 7.36. Draw the outline of the pawl scale 1:1.

Fig. 7.37

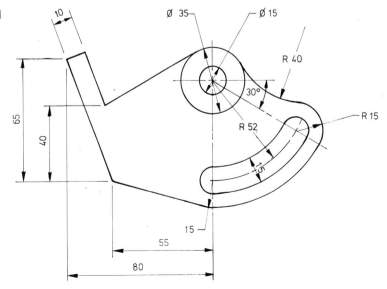

7. The profile of a reclining seat bracket is given in Fig. 7.37. Draw the given view scale 1:1, using geometrical constructions for the tangents and tangential arcs.

8 The use of dry-transfer methods of lettering, shading and toning

An alternative method of enhancing drawings is by the dry-transfer method. Commercially produced sheets of printed transfers are available in a wide range of type faces, symbols and textures and are applied to drawings by simply rubbing the transfer on to the drawing by the method shown in Fig. 8.1.

The last photograph shows a technique whereby letters can be pre-released before fixing to a particularly difficult shape.

Examples of type faces are given in Fig. 8.2

Symbols available include among many others:

borders	arrows
buildings and transport	electrical symbols
architectural symbols	symbols for use in graphs

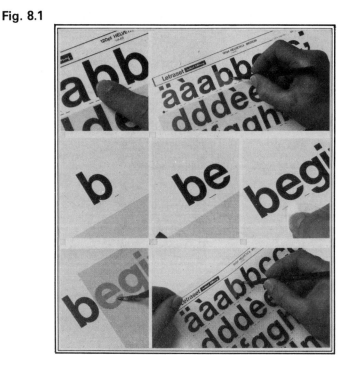

Fig. 8.1

Fig. 8.2

Compacta Bold

COMPACTA BOLD OUTLINE

Futura Medium

Futura Medium Italic

Helvetica Light

Helvetica Light Condensed

PROFIL

Univers 55

Times Extra Bold

Toning and shading transfers are also available. The method of using these is shown in Fig. 8.3.

Fig. 8.3

9 Applied geometry 4

The ellipse

An ellipse is a plane figure produced when a cone is cut obliquely by a plane making a smaller angle with the base than the side of the cone makes with the base, as shown in Fig. 9.1. An example of the cone of an ellipse is shown in Fig. 9.2.

Fig. 9.1

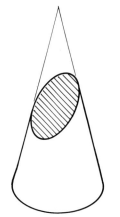

Fig. 9.2

Parts of an ellipse (Fig. 9.3)

Fig. 9.3

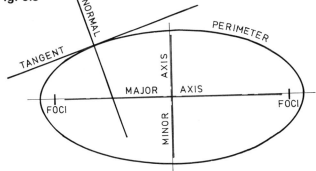

To construct an ellipse – The auxiliary circles method

Step 1 (Fig. 9.4)
Draw the major and minor axes. Note that these two lines bisect each other at 90°.

Fig. 9.4

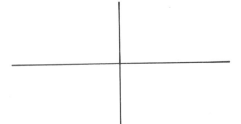

Step 2 (Fig. 9.5)
With centres at the intersection of the major and minor axes, draw two concentric circles the diameters of which are equal to the major and minor axes.

Fig. 9.5

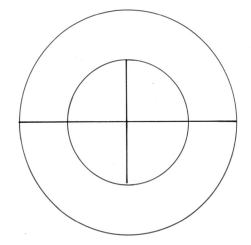

42 The ellipse

Step 3 (Fig. 9.6)
Draw lines through the centre to the circumference of the larger circle. The number of lines drawn will depend upon the size of the circle.

Fig. 9.6

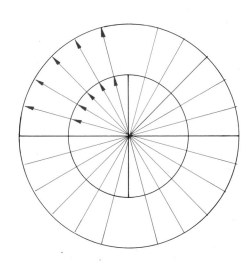

Step 4 (Fig. 9.7)
At the point where each of the radiating lines intersect with the smaller circle, draw a horizontal line to intersect with a vertical line drawn from each of the points where the radiating lines intersect with the larger circle. The points of intersection between the vertical and horizontal lines lie on the ellipse.

Fig. 9.7

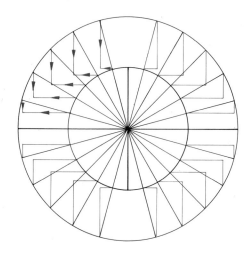

Step 5 (Fig. 9.8)
Join the points of intersection to obtain the ellipse.

Fig. 9.8

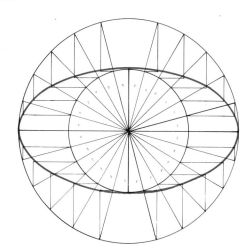

The ellipse

To construct an ellipse – the foci method

Step 1 (Fig. 9.9)
Find the foci of the ellipse by drawing arcs, centre one end of the minor axis, radius half the length of the major axis.

Fig. 9.9

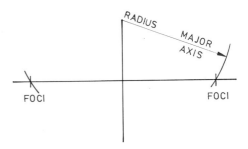

Step 2 (Fig. 9.10)
Mark any point P between the foci.
Draw arcs radius $R1$ and $R2$ to intersect as shown.

Fig. 9.10

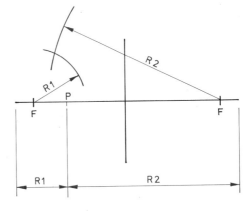

Step 3 (Fig. 9.11)
Repeat step 2 for other points between the foci.

Fig. 9.11

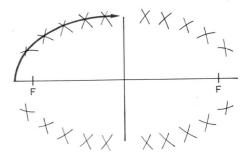

Step 4 (Fig. 9.12)
Join the points of intersection of each pair of arcs to form the ellipse. Aids for the smooth drawing of curves facilitate drawing good quality outlines.

Fig. 9.12

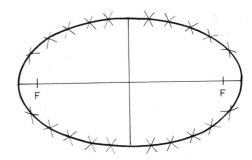

44 The ellipse

To construct a tangent to an ellipse

Step 1 (Fig. 9.13)
Join the required point of tangency P to the foci F1 and F2. Bisect the angle between these two lines to find the normal.

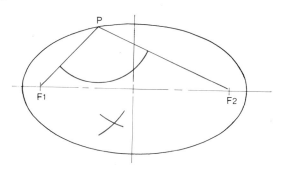
Fig. 9.13

Step 2 (Fig. 9.14)
Construct an angle of 90° to the normal at P to form the tangent.

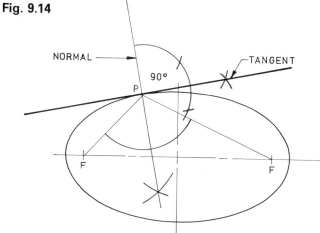
Fig. 9.14

A practical method of drawing an ellipse (Fig. 9.15)
An ellipse can be drawn using a length of string and two pins. Secure the string, which must be equal in length to the major axis of the ellipse, at its ends at the foci of the ellipse. Use a pencil or other marking tool to extend the string and draw the ellipse as shown in the photograph.
This method is used to mark out large ellipses in gardening or building.

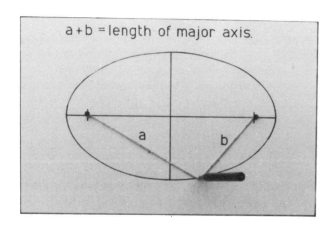
Fig. 9.15

To construct an ellipse – the trammel method (Fig. 9.16)
A trammel is a length of paper or wood on which half the major axis and half the minor axis of the ellipse are marked (Fig. 9.16).

The trammel is moved so that the point x is always on the major axis and the point y is always on the minor axis. Points on the ellipse are then marked as shown.

In drawing ellipses using this method, an excessive amount of construction is avoided, although care must be taken to ensure that dots are sufficiently small to be absorbed in the outline.

The trammel method is also used for practical work where it is not possible to use pins.

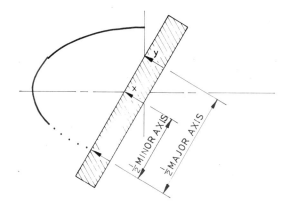
Fig. 9.16

The ellipse – examples

1. The outline of the end of a spanner is given in Fig. 9.17. Draw scale 3:1 (three times full size), the given outline showing all construction lines.

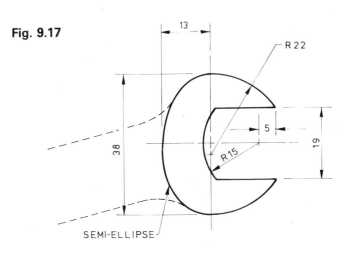

Fig. 9.17

2. The outline and an pictorial representation of a heavy-duty foot pedal are shown in Fig. 9.18. Draw, scale 1:1, an outline of the pedal showing the geometrical constructions for the quarter ellipse and the tangential arcs.

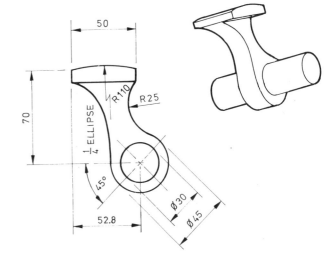

Fig. 9.18

3. The arch of a model bridge is in the form of a semi-ellipse with the major axis 120 mm and the minor axis 70 mm. The spacing between the stones of the arch is 15 mm, i.e. the distances AB, BC and CD are 15 mm. The lines indicating the joints between the stones are normals to the ellipse (Fig. 9.19).
Draw the arch scale 1:1 showing your construction for the semi-ellipse and the three normals at B, C and D.

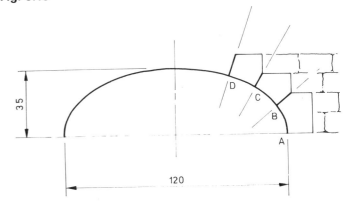

Fig. 9.19

4. Two views of an elliptical cam are given in Fig. 9.20. Do not copy the given views, but draw a scale 1:1 *oblique* view of the cam. Hidden detail is not required.

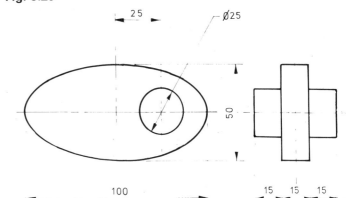

Fig. 9.20

46 The ellipse — examples

Fig. 9.21

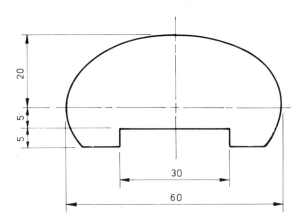

5 The cross-section of an elliptical hardwood handrail is given in Fig. 9.21. The handrail is grooved on the underside to receive the metal wall supports.
Draw the cross section scale 2:1.

Fig. 9.22

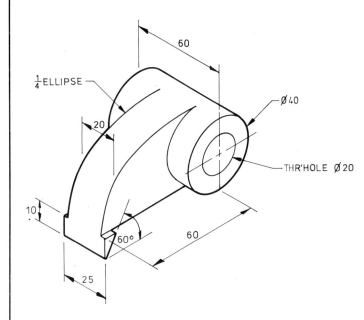

6 A pictorial view of a pawl is given in Fig. 9.22. Do not copy the given view, but instead draw a scale 1:1 oblique view showing the geometrical construction for the quarter-ellipse. Hidden detail is not required.

Fig. 9.23

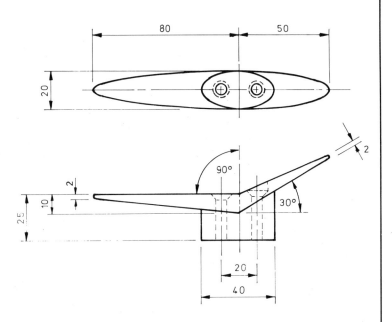

7 Two views in orthographic projection of a halliard cleat are given in Fig. 9.23. Draw scale 1:1, the two views showing clearly the geometrical constructions for the elliptical and semi-elliptical faces. The screw holes may be omitted from your drawing.

Fig. 9.24

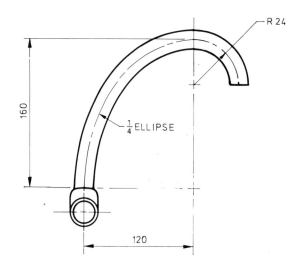

8 The outline of a domestic mixer tap which is formed of 20 mm outside diameter tubing, is shown in Fig. 9.24.
Draw, scale 1:1, the outline of the tube showing clearly the construction for the quarter-ellipse.

10 The use of shading and colour

Examples in the use of shading

Illustrations can be enhanced by the discreet use of shading and colour, although great care must be exercised in using either of these media, and it must be emphasised that excessive shading or colour cannot be used to cover up bad drawings. Before starting this section read page IV of the colour section, following page 48.

Note the effect produced by shading in the examples given in Figs 10.1, 10.2 and 10.3.

Fig. 10.1

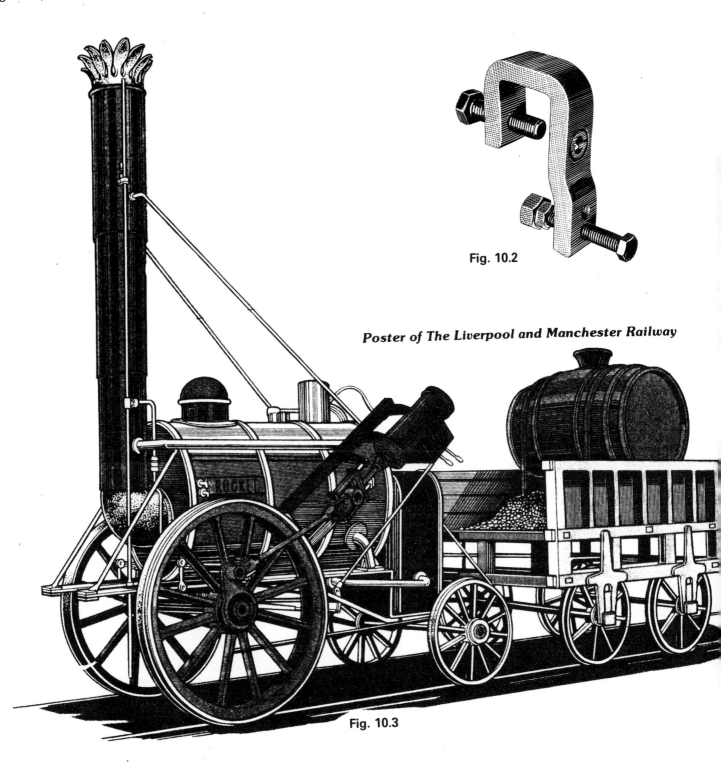

Fig. 10.2

Poster of *The Liverpool and Manchester Railway*

Fig. 10.3

48 Examples for shading and colouring

A4 each

Copy the given drawings twice the size printed and then add colouring or shading most suitable for the given use.

Fig. 10.4

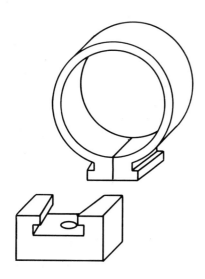

Fig. 10.5

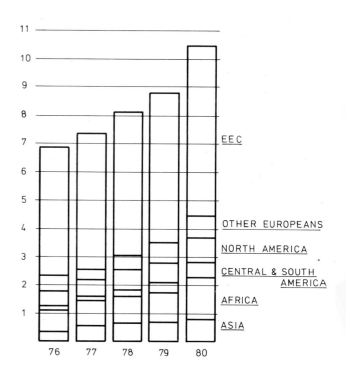

1. An exploded view of the two parts of a plastics pipe clip for a DIY magazine (Fig. 10.4).

2. A graph for a Company Annual Report (Fig. 10.5).

Fig. 10.6

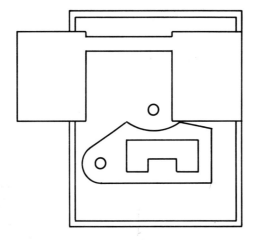

3. The working parts of a desk lock for an Instruction Manual (Fig. 10.6).

Fig. 10.7

4. An illustration of a new house for a House Agent's Brochure (Fig. 10.7).

Materials for shading and colour

Pencils, crayons and other point media

Pencils can be obtained in various degrees of hardness (H) and blackness (B). The range from H to 2B is the most useful for shading, (Fig. 10.11).

Clutch pencils avoid the need for sharpening, (Fig. 10.12).

Crayons can be obtained in various thicknesses and colours. The thinner variety are the most useful as they have a greater degree of accuracy, (Fig. 10.13).

Fig. 10.11

Fig. 10.12

Fig. 10.13

Pens and markers

Fig. 10.14

Ball-point pens give constant line width, (Fig. 10.14).

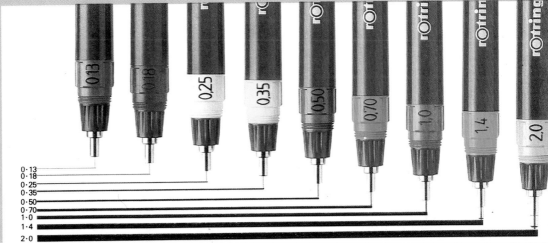

Fig. 10.15

Draughting pens give a high degree of accuracy; interchangeable nibs give a wide range of line thicknesses, (Fig. 10.15).

Felt-tip markers give greater coverage, (Fig. 10.16).

Fibre-tip pens can give a greater variety of line width (Fig. 10.17).

Fig. 10.18

Plastic-tip pens give uniform ink flow, (Fig. 10.18).

Brushes for use with ink or paint

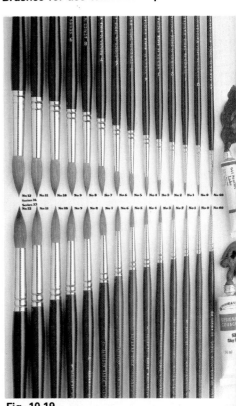

Good quality brushes should be used with drawing or Indian inks, water colours or process colours, (Fig. 10.19).

Shading can also be applied using dry-transfer methods, see page 40.

Fig. 10.19

The use of colour

The use of colour, applied using the materials and techniques described on the previous page, is another method of enhancing an illustration.

Some of the uses of colour are shown below and on the next page. It must be emphasised that when colour is used, it must be applied with great care.

Fig. 10.20

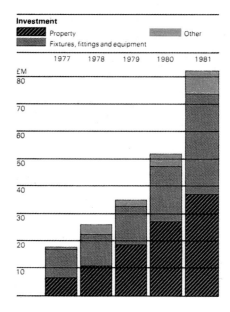

To differentiate between areas, i.e. as in the graph Fig. 10.20. Note that in this illustration, the colours tone in with each other.

To accentuate or emphasise parts of an illustration as in the pictogram Fig. 10.21. Notice how a bright colour has been used for the largest area.

Fig. 10.21

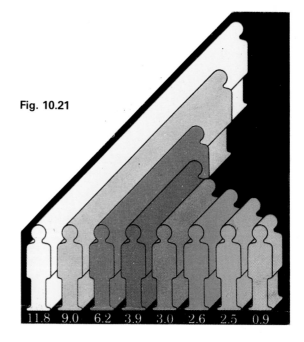

To colour-code parts of a diagram in order to refer to additional written information, Fig. 10.22. Here the colours must be clearly different.

Fig. 10.22

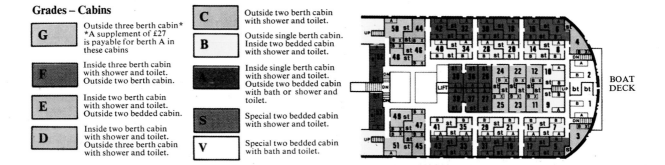

The use of colour

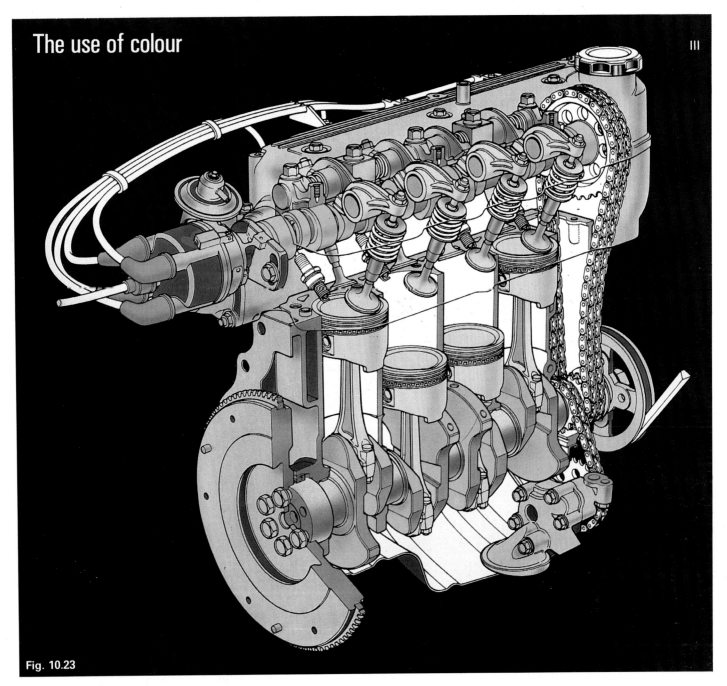

Fig. 10.23

To distinguish between parts of an illustration in order to be able to describe them, i.e. in the illustration of the Mazda 323 engine (Fig. 10.23).

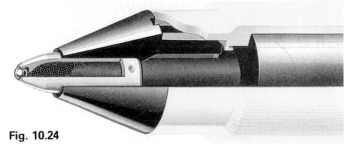

Fig. 10.24

As a means of shading to give greater realism, as in the cut-away illustration of the ball pen (Fig. 10.24).

To pick out parts of a map (Fig. 10.25).

IV The use of line and shading

Line thickness is an important aspect of technical illustration. The effect produced by different line thicknesses on a cube is illustrated below.

Constructed objects are often built-up of geometrical solids; the two most common of which are the prism and the cylinder. Methods of shading these two solids are illustrated below. Notice the effects produced in J and Q to give the impression of wood and in L and S to give the impression of glass. Which of the other illustrations gives an indication that the solid is made of polished metal?

A All these lines are the same thickness, and, as a result, the figure could be anything depending on the eye of the viewer (Fig. 10.26).

Fig. 10.26

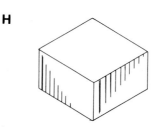
H

O

B With some lines thicker than others, the figure adopts a more solid form which, in this illustration, is obviously viewed from above (Fig. 10.27).

Fig. 10.27

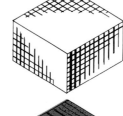
I

P

C This is like Fig. 10.27 but now the object is viewed from below (Fig. 10.28).

Fig. 10.28

J

Q

D An even more solid form is created by tapering some of the lines (Fig. 10.29).

Fig. 10.29

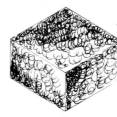
K

R

E This is like Fig. 10.29 but viewed from below (Fig. 10.30).

Fig. 10.30

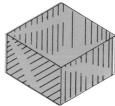
L

S

F In this figure, you could be looking at a cube or into a corner as all the lines are the same thickness (Fig. 10.31).

Fig. 10.31

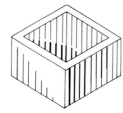
M

T

G With some of the lines tapered, the object is clearly a cube (Fig. 10.32).

Fig. 10.32

N

U

11 Graphs

Examples of six three-dimensional graphs are given below. Although each of them is suitable for many uses, some, obviously, are more suited to one use than another.

Fig. 11.1

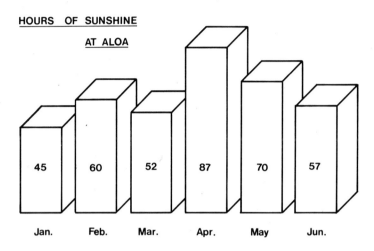

Block graph
Used to represent quantities on a periodic basis; colour could be used to improve this graph (Fig. 11.1).

Fig. 11.3

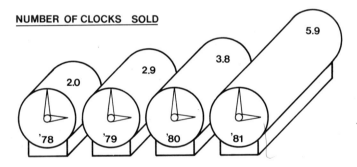

Horizontal block graph
Used to represent quantities on a periodic basis. Notice the use of an ideogram to enhance the graph (Fig. 11.3). The position of the hands could also be used to represent the amounts in this particular example.

Fig. 11.5

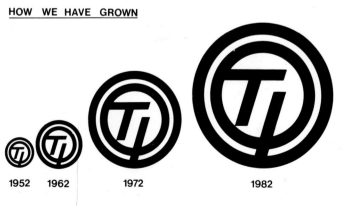

A graph
Growth over a period is represented by increasing the size of a logo (Fig. 11.5).

Fig. 11.2

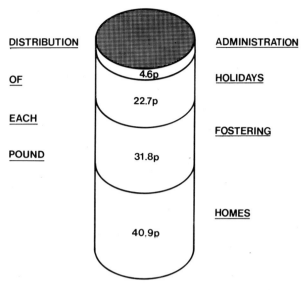

Cylindrical pie graph
Used to show the distribution of a fixed amount between a number of parts (Fig. 11.2).

Fig. 11.4

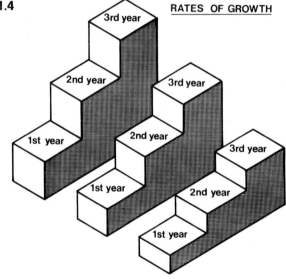

A comparative block graph
Used to compare different rates of growth (Fig. 11.4).

Fig. 11.6

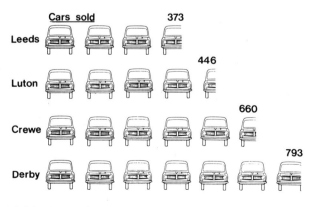

A block graph
Numbers are represented by pictorial units (Fig. 11.6).

Graphs – examples

1 A wine seller wishes to compare the different ways in which the cost of a bottle of wine is distributed according to the price. The figures are as follows:

Cost of bottle of wine	Percentage				
	duty	bottling	margin	VAT	wine
£1·50	31	12	30	15	12
£2·00	25·5	10	28	15	21·5
£2·50	20	8	27	15	30

Draw a graph to illustrate these figures.

2 The relative percentages of freight moved by various methods in 1981 was by:

road 64%

rail 12%

ship 18%

pipeline 6%

Illustrate these figures pictorially.

3 The percentage of vehicles surviving in 1981, according to the year the vehicles were first registered, is as follows:

Year of registration	Percentage of vehicles surviving
68/69	30
70/71	51
72/73	74
74/75	87
76/77	93
78/79	94
80/81	97

Draw a graph to illustrate these figures.

4 A company selling a Family Bond wishes to illustrate its growth in comparison with other forms of saving.

The figures to be illustrated are as follows:

Value of savings for each £20 invested monthly in four different types of scheme.

	After 5 years	After 10 years	After 15 years
Building Society	3100	5800	8600
Life Assurance	4800	7200	9900
Index-linked SAYE	5000	9000	16 400
Family Bond	8000	20 000	54 000

Draw a suitable graph to illustrate these figures.

A further example of a three-dimensional graph is given in Fig. 11.7.

Fig. 11.7

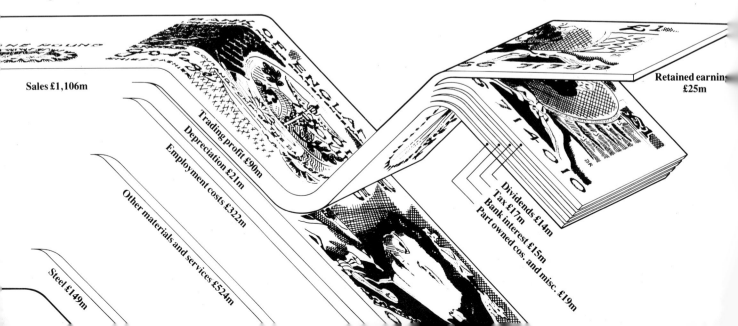

Sales £1,106m · Trading profit £90m · Depreciation £21m · Employment costs £322m · Other materials and services £524m · Steel £149m · Retained earnings £25m · Dividends £14m · Tax £17m · Bank interest £15m · Part owned cos. and misc. £19m

12 Applied geometry 5   Loci

The word locus means place; in geometry it is used to mean path.
To take examples of loci from a pendulum clock

The path or locus of a weight of the clock as it moves downwards is a straight line (Fig. 12.1).

The locus of the pendulum as it swings is an arc (Fig. 12.2).

The locus of the end of one of the hands is a circle (Fig. 12.3).

Fig. 12.1

Fig. 12.2

Fig. 12.3

Fig. 12.4

The locus of the *centre* of a wheel rolling along a flat surface is a straight line, parallel to the flat surface and the radius of the wheel above it (Fig. 12.4).

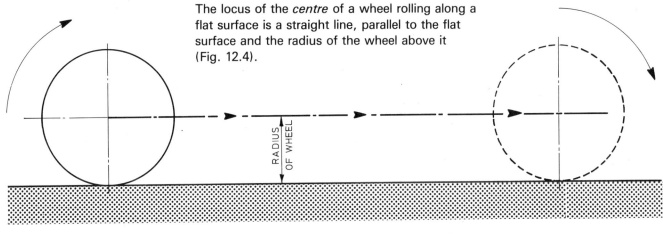

Fig. 12.5

When the wheel reaches a right-angled corner, the locus of the centre of the wheel is an arc, the radius of which is the same as the radius of the wheel (Fig. 12.5).

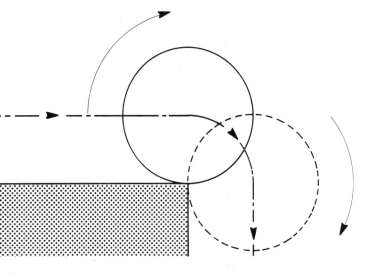

52 Loci

A pictorial view of a motor cycle engine is given in Fig. 12.6.

A sectional drawing through the same engine is given in Fig. 12.7.

This is an example of a construction in which the materials used, i.e. various metal alloys, have affected the shape, dimensions and proportions of the parts.

As the piston goes up and down in the cylinder, the crank is turned by the connecting rod. Notice that the locus of the point A of the piston is a straight, vertical line, and the locus of the point B of the connecting rod is a circle.

In the design of engines or similar mechanisms, it may be necessary to determine the locus of other moving points, for example the point C in the centre of the connecting rod. This can be done by drawing a diagram of the mechanism as shown in Fig. 12.8 in which a number of positions of the point are found and then joined up to form the locus.

In the drawing, twelve positions on the locus have been found. It is easy to divide a circle into twelve equal divisions using a 60°/30° set square. Twelve points are sufficient to draw an accurate locus.

Aids for the smooth drawing of curves such as french curves, flexible curves and ellipse guides can be used to obtain a good quality outline for loci.

Fig. 12.6

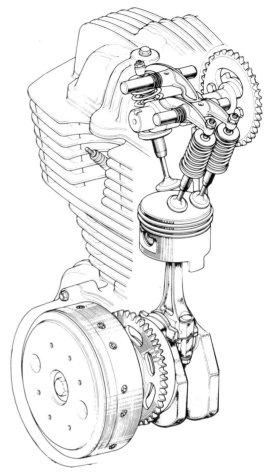

Fig. 12.7

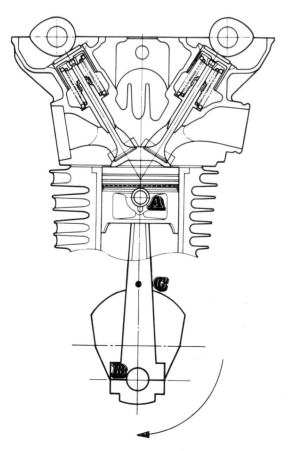

Fig. 12.8

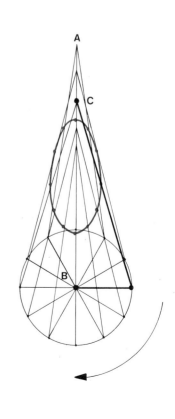

Loci

Mechanisms similar to the one shown on the previous page are given below.

Fig. 12.9

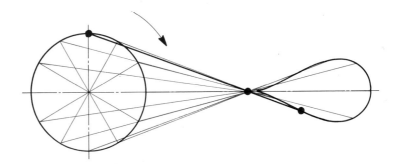

Mechanism with a link constrained to pass through a fixed point (Fig. 12.9)

Construction:

Divide the locus of the revolving end (the circle) into twelve equi-spaced positions.

From these points, draw twelve positions of the link.

Join the ends of the link to construct the locus.

Fig. 12.10

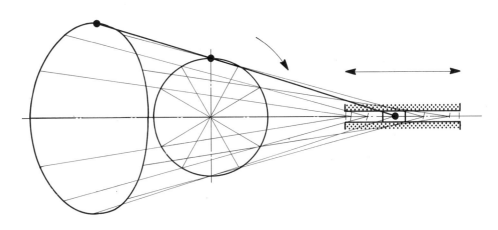

Mechanism with the end of the link constrained to move in a horizontal line (Fig. 12.10)

Construction:

Divide the locus of the revolving end (the circle) into twelve equi-spaced positions.

From these points, using compasses, draw twelve positions of the link.

Mark the points required for the locus on each of the twelve positions of the link.

Join these points to construct the locus.

Fig. 12.11

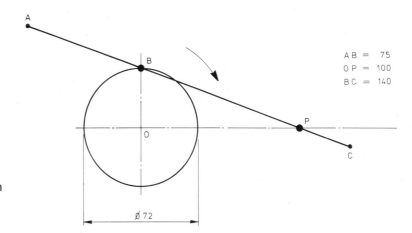

```
AB = 75
OP = 100
BC = 140
```

Figure 12.11 is a diagram of another mechanism.

Draw the view to the dimensions given, and then construct the locus of points A and C for one revolution of B about O.

Loci

A3 each

Fig. 12.12

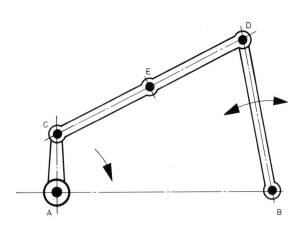

Fig. 12.13

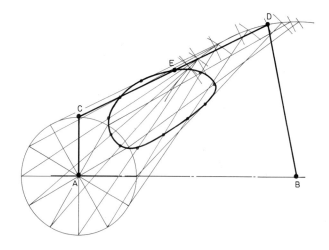

A link mechanism comprising three links is shown in the sketch Fig. 12.12. AC revolves about A, and DB oscillates about B. C and D are pin-jointed.

A diagram of the same mechanism is given in Fig. 12.13.

Construction for drawing the locus of *E*:

Divide the circle, centre A, into twelve equal parts.

Draw the arc radius DB centre B.

Using compasses, mark off CD in each of the twelve positions.

Mark the point E on each line.

Join up the points to construct the required locus.

Fig. 12.14

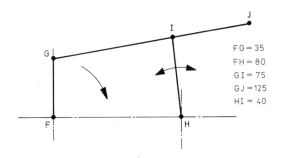

FG = 35
FH = 80
GI = 75
GJ = 125
HI = 40

Fig. 12.15

KL = 30
MN = 30
LN = 120
KM = 120
LO = 40

Draw the locus of J for one revolution of G about F (Fig. 12.14).

Draw the locus of O for one revolution of L about K (Fig. 12.15).

Fig. 12.16

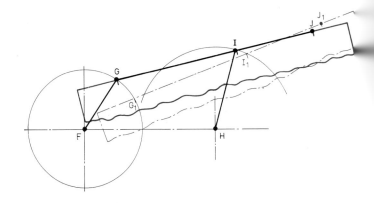

Using a trammel

An alternative method for drawing a locus is by using a strip of paper, called a trammel on which the appropriate distances are marked. The use of a trammel to draw the locus required in Fig. 12.15, is shown in Fig. 12.16.

The trammel is placed on the drawing in each of the twelve positions of G and I in turn, and points J are marked with a dot on the drawing. The locus is then drawn in.

Loci – examples

1. The linkage of a mechanically operated lever system is shown in the sketch Fig. 12.17, and the diagram Fig. 12.18. When the handle A is in the extreme left-hand position, the pin joint at B is at C and the pin joint D is at E. When the handle is in the extreme right-hand position, the pin joint B moves to E.
   Draw the given diagram and plot the locus of P on the line BD for the complete movement of the handle.

2. The sketch, Fig. 12.19, shows part of a wrapping machine. Fig. 12.20 is a line drawing of the same mechanism. The mechanism consists of a crank ED connected by a pin joint to a rod EF. As the crank rotates, the end of the rod F slides along the quarter elliptical slot BC. Given that AC = 70, AB = 52, AD = 84, plot the locus of G for one complete revolution of the crank.

3. Both parts of this question are concerned with the drop front of a cupboard
   (a) Figure 12.21 shows the drop front fitted with a hinged stay ABC, which allows the front to drop only to the horizontal position.
   Draw the locus of B as the drop front closes from the horizontal to the vertical position.

   (b) Figure 12.22 shows the drop front fitted with a sliding stay DE, which also allows the front to drop only to the horizontal position. The stay pivots at D and slides along a pin at F. Draw the locus of E as the front closes from the horizontal to the vertical position.

   (c) What other materials could be used for the stay? If plastic or nylon cord were used instead of the metal shown, how would this affect the shape of the parts? Draw sketches to illustrate the changes.

4. Part of the feed mechanism of a machine is shown in the sketch Fig. 12.23. A line drawing of the mechanism is shown in Fig. 12.24. The rod KL oscillates about the pivot K while the end M of the rod LM slides horizontally.
   Draw the view Fig. 12.24, and plot the locus of N the mid point of LM as M moves from M_1 to M_2.

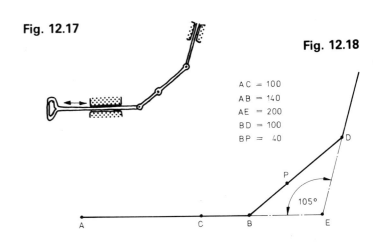
Fig. 12.17

Fig. 12.18

AC = 100
AB = 140
AE = 200
BD = 100
BP = 40

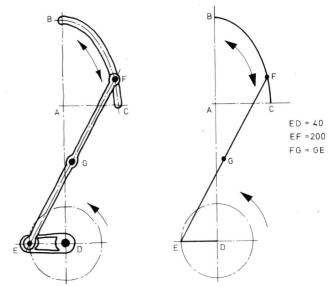
Fig. 12.19   Fig. 12.20

ED = 40
EF = 200
FG = GE

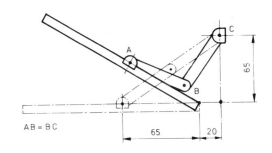
Fig. 12.21

AB = BC

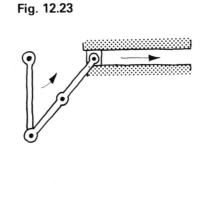
Fig. 12.22

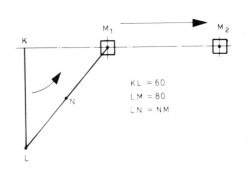
Fig. 12.23   Fig. 12.24

KL = 60
LM = 80
LN = NM

56 The cycloid

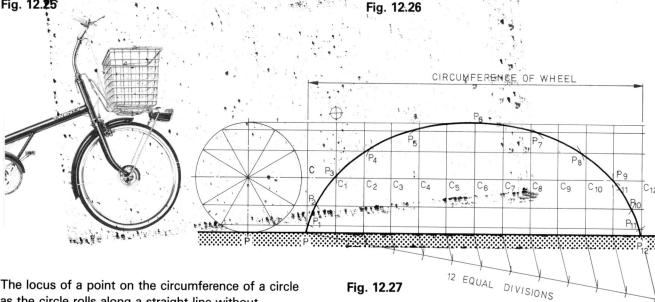

Fig. 12.25

Fig. 12.26

The locus of a point on the circumference of a circle as the circle rolls along a straight line without slipping is known as a cycloid. The photograph, Fig. 12.25, gives a practical illustration of a cycloid.

Construction (Fig. 12.26):

Divide the circle into twelve equal parts.
Draw the line $P-P_{12}$ in projection with the base of the circle. Make the length of the line equal to the circumference of the circle. This may be done in two ways:
– calculate the circumference using πD;
– step off the length of one of the divisions in the circle twelve times along the line.

If the calculation method is used, divide the distance into twelve equal parts (see page 17).
Draw vertical lines at each division.
Project horizontal lines from each of the twelve divisions in the circle. With centres C, C_1, C_2, etc., draw sufficient of each circle to establish points P_1, P_2, etc.
Join the points P to construct the cycloid.

Examples

1. Construct the locus of a point on a wheel diameter 63 mm as it rolls along a line without slipping for one complete revolution of the wheel.

2. A wheel diameter 80 mm rolls along the line AB without slipping (Fig. 12.27).
   Draw the locus of P from the given position until it is in contact with the line.

3. A wheel diameter 50 mm rolls along the lines $ABCD$ without slipping (Fig. 12.28).
   Draw the locus of P as the wheel rolls from A to D.

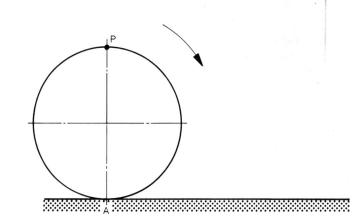

Fig. 12.27

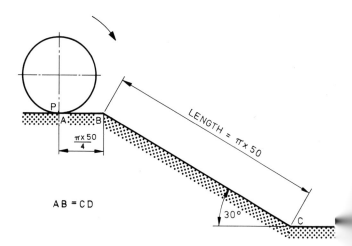

Fig. 12.28

The helix

Fig. 12.30

The locus of a point which moves at a uniform speed concurrently along and round the periphery of a cylinder is known as a helix. The distance travelled along the cylinder during one complete revolution of the point round the cylinder is the pitch of the helix.

An example of a helix is given in the photograph Fig. 12.29, and the elevation Fig. 12.30.

Fig. 12.29

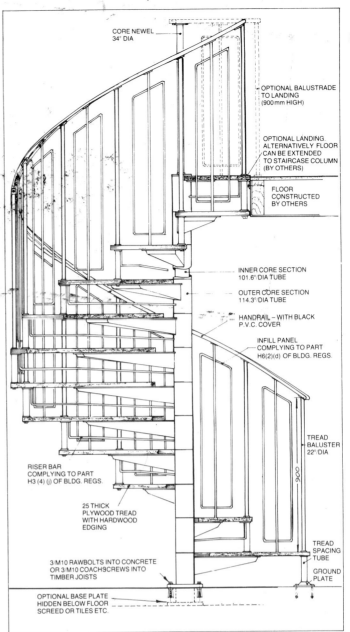

Fig. 12.31

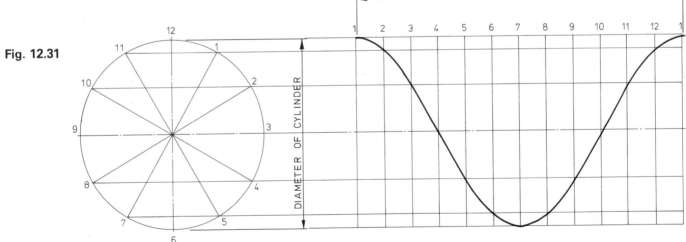

To construct a helix:

Draw an end view, i.e. a circle, and a side view, i.e. a rectangle, of the cylinder.

Divide the circle into twelve equal parts using a 30°/60° set square.

Mark the pitch of the helix in position on the cylinder and divide it into twelve equal parts.

Project the divisions in the end view to the relevant lines in the side view.

Draw the helix through the points as indicated in Fig. 12.31.

You may find it helpful to number the divisions in the end view and side view as shown.

Example

Draw a helix with a pitch of 180 mm for two turns round a cylinder of 100 mm diameter.

58 Loci – examples

A3 each

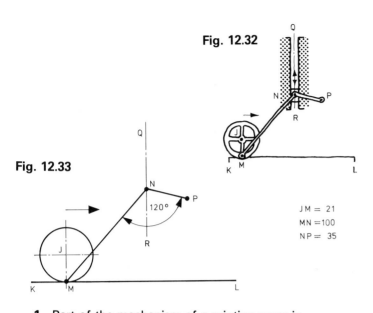

Fig. 12.32

Fig. 12.33

JM = 21
MN = 100
NP = 35

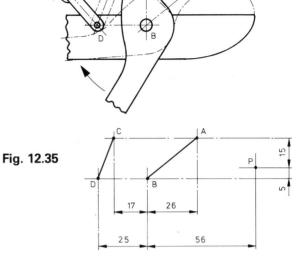

Fig. 12.34

Fig. 12.35

1 Part of the mechanism of a printing press is shown in the sketch Fig. 12.32. A diagram of the mechanism, reduced to a suitable scale, is given in Fig. 12.33.

As the wheel, centre J, rolls without slipping along the horizontal surface KL, the slider at point N of the rod MNP moves vertically up and down the centre line QR. The line QR is vertically above the highest point reached by M.

Plot the locus of P as the wheel rolls along the surface KL until M reaches its highest point.

2 The jaws of a pair of secateurs are shown in Fig. 12.34. The jaws, fully closed, are indicated by the short chain lines. A diagram of the mechanism with the jaws fully closed is given in Fig. 12.35.

Copy this diagram, scale 2:1, and then plot the locus of the end P as the jaws open from the fully closed position until the line AB is vertical.

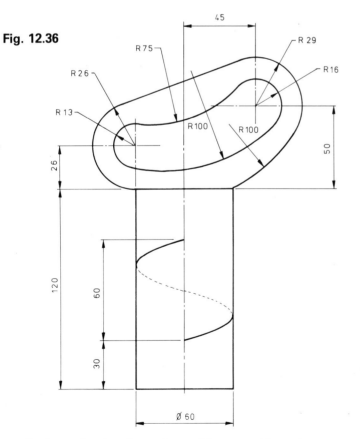

Fig. 12.36

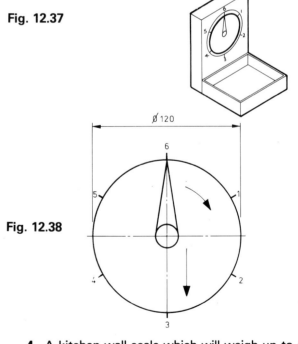

Fig. 12.37

Fig. 12.38

3 A starting key for a piece of farm machinery is shown in Fig. 12.36. It comprises a flat handle as shown and a cylindrical block of metal around which a narrow helical slot is cut.

Draw the given view scale 1:1. The helical slot may be represented by a single line. Hidden detail is required.

4 A kitchen wall scale which will weigh up to a maximum of 6 kg is shown in Fig. 12.37. The scale is attached to the wall by a separate back plate. When an object is placed in the pan, the entire scale moves downwards and the weight of the object is recorded by the pointer on the dial. When the scale is used to weigh 6 kg, the pointer makes one complete revolution of the dial and the scale moves downwards 70 mm. Copy the simplified drawing of the dial (Fig. 12.38) and plot the locus of the end P of the pointer as the scale is used to weigh 6 kg.

13 Applied geometry 6 — Development – prisms

Fig. 13.1

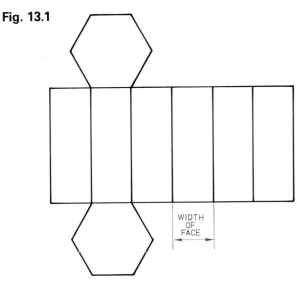

The development of a solid is the surface of that solid unfolded or unrolled and laid out on a plane (Fig. 13.1).

Development of a prism (Fig. 13.2)

Draw a plan and front view of the prism.

Unfold the six sides as shown.

Add the top and bottom in a convenient position.

Fig. 13.2

Fig. 13.3

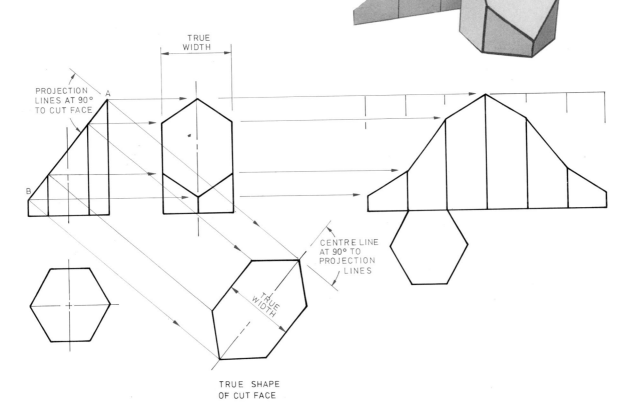

Development of a truncated prism (Fig. 13.3)

Draw the development of the prism without truncating it.

Draw the cutting plane line *AB*.

Project the points at which the cutting plane line intersects with the vertical edges of the prism to the development.

Line in the completed development.

Draw an auxiliary view to find the true shape of the inclined face.

Development of prisms – examples

A3 each

Fig. 13.4

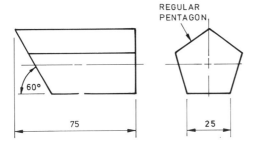

1 Two views of the engine compartment of a model tractor are given in Fig. 13.4.

Draw the given views scale 1:1.

Project the plan view.

Draw the development of the surface of the engine compartment including the two ends.

See page 67 for further work on this project.

2 Two views of a box for a chocolate easter egg are given in Fig. 13.5.

Draw the two views scale 1:1, and from them project the surface development.
If you have time, make a model of the box and decorate it.
What other material could be used instead of cardboard for the box? How would this affect the design?

Fig. 13.5

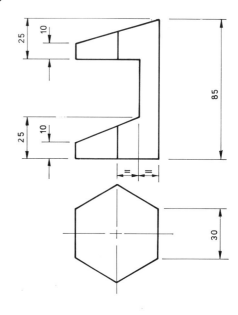

Fig. 13.6

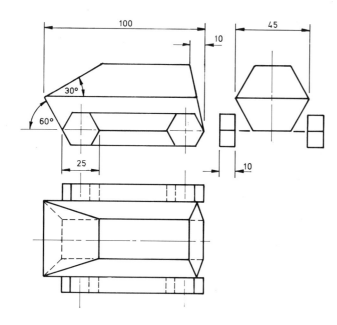

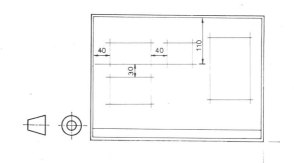

3 Orthographic views of a model tracked snowcat are given in Fig. 13.6.

Draw the given views scale 1:1, and from them project a development of the surface of the component parts separately.
This is another model you can make and decorate if you have time.

Development – cylinders

Fig. 13.7

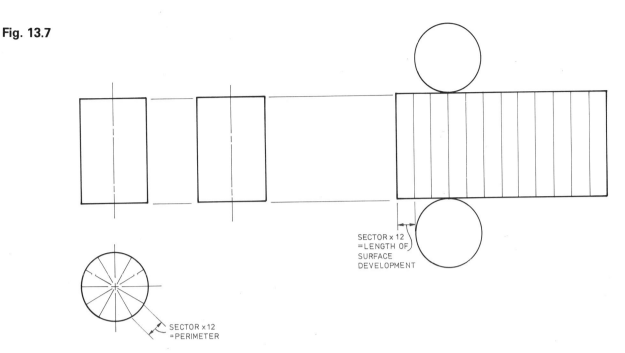

Development of a cylinder (Fig. 13.7)

A cylinder is developed by unrolling.

Draw the plan and front view of the cylinder.

Project the height of the development from the front view.

Determine the length of the development which is the same as the circumference of the cylinder by:
either – calculation using example given in Fig. 13.5;
or – dividing the plan into twelve equal parts and stepping off the twelve divisions along the development.

Draw the top and base in an appropriate position.

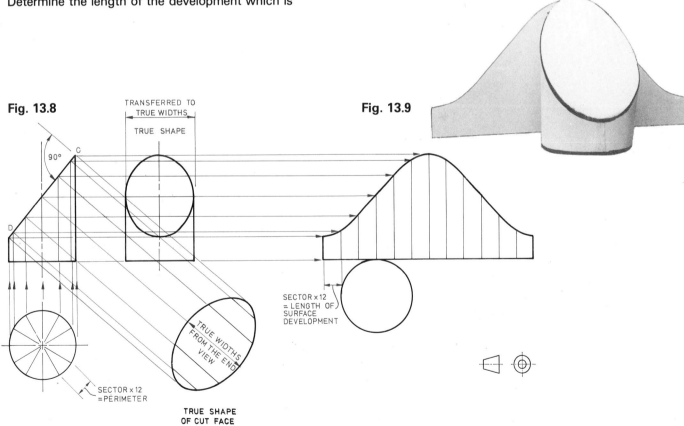

Development of a truncated cylinder (Fig. 13.8)

Draw the development of the cylinder without truncating it.

Draw the cutting plane *CD*.

Draw the ordinates on the front view and the development.

Project the points where the cutting plane intersects with the ordinates on the front view to the similar ordinates on the development.

Join the points to complete the development as shown (Fig. 13.9).

Draw an auxiliary view to find the true shape of the inclined face.

Development of a cylinder – examples

A3 each

1. You are required to make a model of the scoop shown in the orthographic views (Fig. 13.10). Draw the two given views scale 1:1, and from them:

   (a) project the plan view;
   (b) construct the development of the surface.

   Include in your drawing the construction for the tangential arcs of the handle.
   The model can be made from good quality thin card and sprayed with paint to strengthen the material.

Fig. 13.10

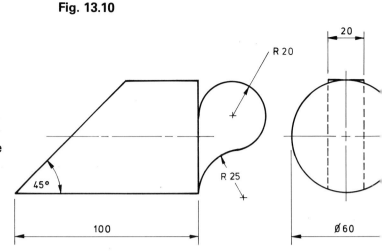

2. A sketch of a candle holder with two views of the semicircular base are given in Figs 13.11 and 13.12.
   Draw, scale 1:1, the following views of the base:

   (a) the given front view;
   (b) the given end view;
   (c) the plan;
   (d) the surface development.

   The development can be used as a pattern to mark out the shape of the base of the candle holder on thin metal. To complete the model, add a pin to support the candle and a circular dish to catch the molten wax.

Fig. 13.11

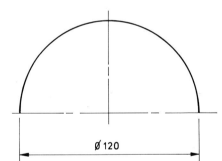

Fig. 13.12

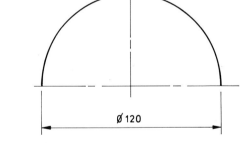

3. A metering instrument, 80 mm ⌀ is to be inserted in the control panel of a machine in such a way that it is in direct line with the operator. To achieve this a cylindrical extension has to be produced to carry the instrument, details of which are given in Fig. 13.13. Make a scale 1:1 development of the extension piece together with a pattern for the hole in the control panel.

Fig. 13.13

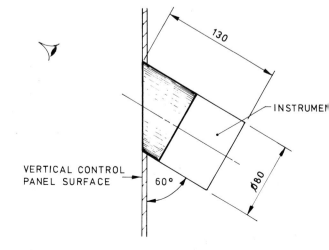

Development – pyramids

Fig. 13.14

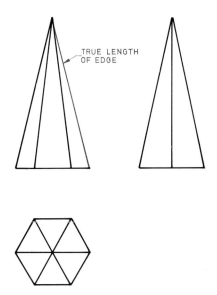

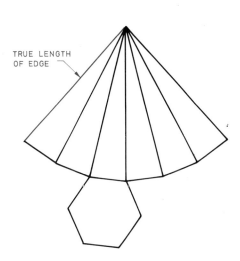

Development of a hexagonal pyramid (Fig. 13.14)

The surface development of a pyramid is radiated about a point. It is important that the radius used is the true length of an edge of the pyramid.

Draw the plan and front view of the pyramid, making sure that the front view includes the true length of the edge.

Draw an arc radius the true length of the edge.

Mark off the length of each side of the base six times along the arc.

Join these points to the centre of the arc to complete the development of the sides.

Draw the hexagonal base in a convenient position.

Fig. 13.15

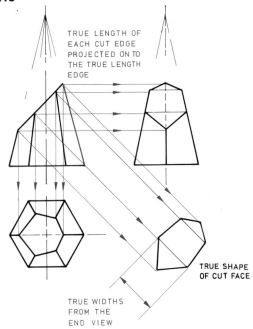

Fig. 13.16

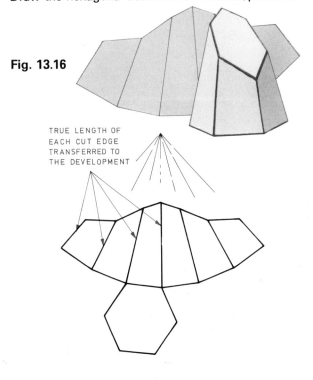

Development of a truncated pyramid (Fig. 13.15)

Draw the development of the pyramid without truncating it.

Transfer the *true length* of each cut edge from the front view to the development.

Join up to complete the development of the sides. Draw an auxiliary view to find the true shape of the inclined face as shown (Fig. 13.16).

64 Development of pyramids – examples

1 A dimensioned, pictorial view of a salt cellar, which is in the shape of a hexagonal-based pyramid truncated at 45°, is shown in Fig. 13.17. Draw, scale 1:1, in either first or third angle orthographic projection, three views of the salt cellar, and from them produce the surface development of the sides, base and inclined top. Could a salt cellar, similar in shape to the one shown, be made of plastic or wood? If so, how would this affect the design?

Fig. 13.17

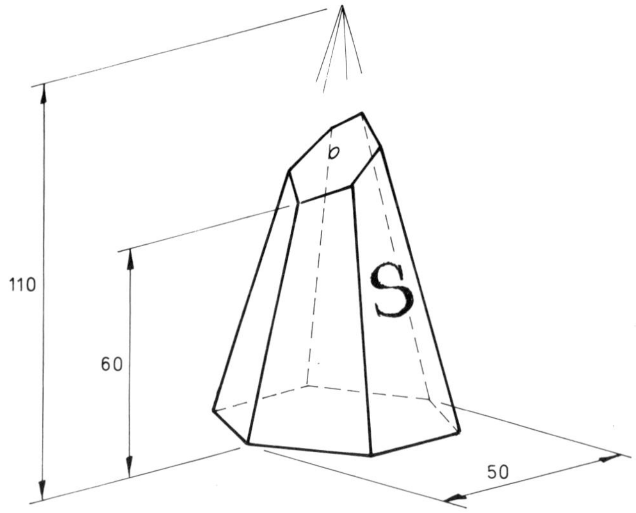

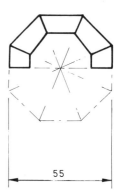

2 The bonnet of a model veteran car is in the shape of half of a truncated octagonal pyramid as shown in Figs 13.18 and 13.19.
Draw the two given views scale 1:1 and project a plan. From these views construct a surface development of the bonnet.
If you wish to make the complete model of the veteran car refer to page 75.

Fig. 13.18

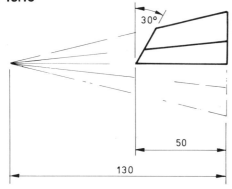

Fig. 13.19

3 A small decorative lantern is shown in Fig. 13.20. The roof of the lantern is in the shape of an octagonal-based pyramid; vertical height 40 mm and distance across the flats of the octagonal base 90 mm.
Draw a scale 1:1 development of the roof only of the lantern. If you wish to make the entire lantern refer to page 69.

Fig. 13.20

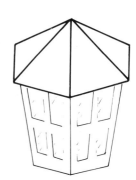

Development – cones

Fig. 13.21

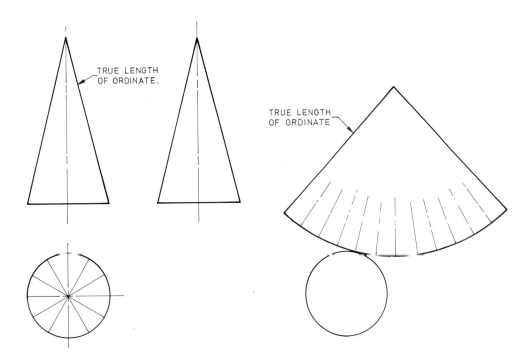

Development of a cone (Fig. 13.21)

As with a pyramid, the surface development of a cone is radiated about a point.
Draw the plan and side view of the cone.
Divide the plan into twelve equal parts using a 30°/60° set square.
Draw an arc radius the true length of an ordinate as shown.

Mark off distances equal to the twelve divisions along the arc.
Join the two end marks to the centre of the arc to complete the development of the curved surface of the cone.
Add the circular base in a convenient position.

Fig. 13.22

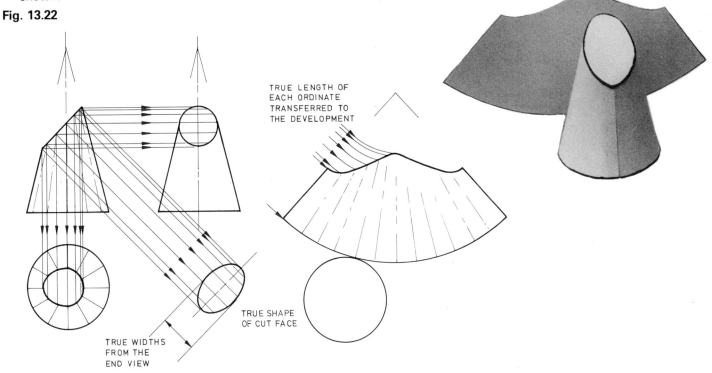

Development of a cone truncated at an angle (Fig. 13.22)

Draw the development of the cone without truncating it.
Draw the ordinates on the plan, from view and development as shown.
Find the *true length* of each ordinate by projecting the point of intersection between the ordinate and the cutting plane to the edge of the cone.

Transfer the true length of each ordinate to the development.
Join up the points to complete the development of the curved surface.
Draw an auxiliary view to find the true shape of the inclined face as shown.

Development of cones – examples

1. The front view of a conical lamp shade is given in Fig. 13.23.
   Draw, scale 1:1, in orthographic projection, the front, end and plan views of the lamp shade and from them produce the surface development.

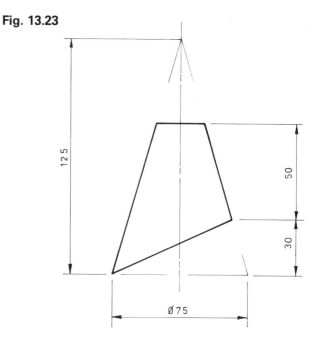

Fig. 13.23

2. A pan for a child's kitchen scales is in the shape of half a truncated cone as shown in Fig. 13.24. Draw the given view scale 1:1, and from that produce the surface development of the pan which is open at the top.

Fig. 13.24

3. To help solve the litter problem at the school tuck shop, it is decided to design, make and install litter bins which are to be attached to the walls of the building. One design is to construct the bins in the shape of a truncated cone as shown in Fig. 13.25.

   (a) Draw the development of the given design scale 1:5.
   (b) Design your own litter bin suitable for the same purpose.
   (c) Decide upon an alternative material for a wall-mounted litter bin and then draw sketches showing how the new material has affected the proportions and construction of it.

Fig. 13.25

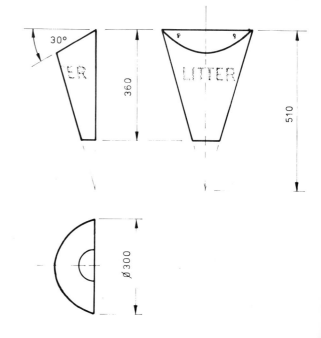

14 Projects   Tractor project

Fig. 14.1

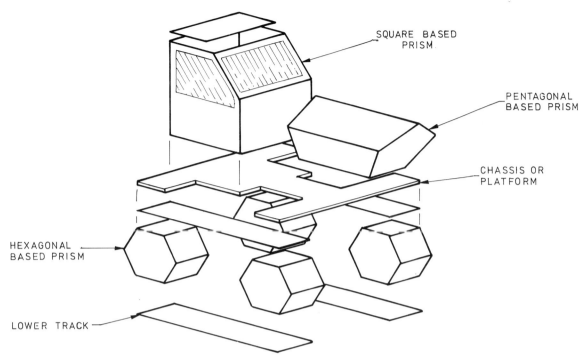

Fig. 14.2

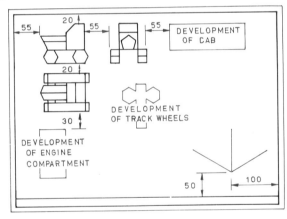

Here is a model that can be made of thin card (Fig. 14.1).

Before the model can be made, it is necessary to make working drawings to include:

(a) orthographic views of the assembled tractor;
(b) a development for each component part.

The dimensions for each part are included in the following drawings.

A suggested layout for the drawings is given in Fig. 14.2, but you may decide on a better layout for yourself.

Fig. 14.3

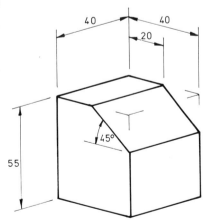

The driver's cab is in the shape of a square-based prism cut by an inclined plane as shown in Fig. 14.3. The development of this part can be projected from the orthographic views as shown in the layout diagram Fig. 14.2.

Tractor project

The engine compartment is in the shape of a truncated pentagonal-based prism, as shown in Fig. 14.4.

Fig. 14.4

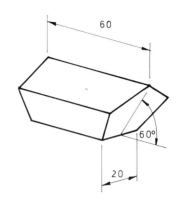

The chassis, or platform, on which the model is built, is made of thick card cut to the shape shown in Fig. 14.5. The overall length and width are given, and the remaining dimensions can be obtained from the base of the driving cab and engine compartment which fit on it.

Fig. 14.5

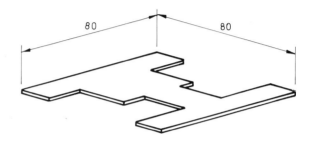

Each of the four wheels are hexagonal prisms as shown in Fig. 14.6.

Fig. 14.6

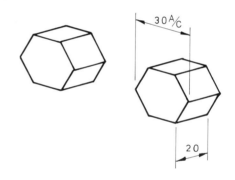

When the model is being made, the development of each part can be traced or pricked through from the drawings on to the material being used.

Complete your model by adding further detail as suggested in the photographs Figs. 14.7 and 14.8 and by colouring it.

Fig. 14.7

Fig. 14.8

Lantern project

Here is another model which can be made of thin card.

To make the lantern, you will need to draw the development of the body and the roof, one of which is a pyramid and the other a truncated pyramid, dimensions of which are given in the orthographic views Fig. 14.9.

A completed model of the lantern is shown in Fig. 14.10.

Fig. 14.10

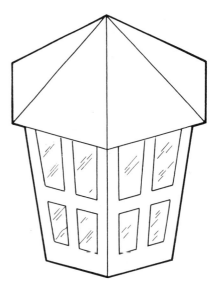

Fig. 14.9

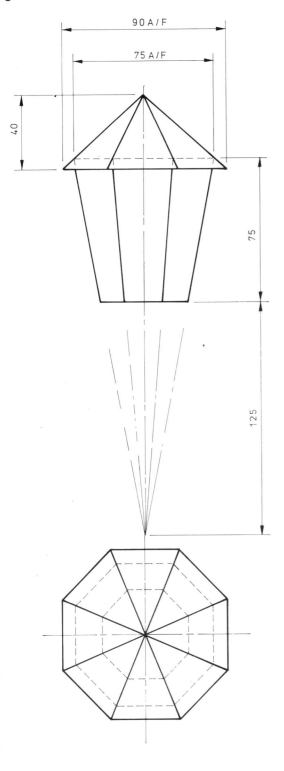

Fig. 14.11

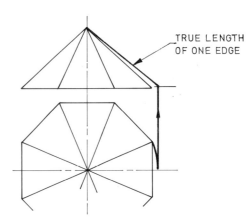

Before the developments can be drawn, it will be necessary to find the *true length* of an edge of the two pyramids. Figure 14.11 shows how the true length of an edge of the roof can be found.

Fig. 14.12

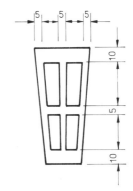

The model can be improved by adding windows. Suggested sizes are shown in Fig. 14.12.

The model can be scaled up and made of sheet metal to produce a porch lamp.

Gemini project

Fig. 14.13

Fig. 14.14

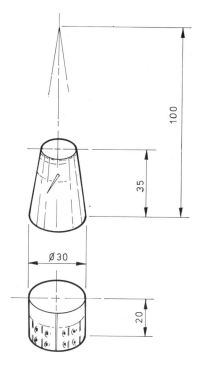

The Gemini space project was developed from knowledge gained from the single-seater Mercury spacecraft. The operation took its name from the Greek word *gemini*, meaning twin, each flight carrying two astronauts.

The first flight took place in March 1965.

The photograph, Fig. 14.13, shows the splashdown in the Pacific Ocean after the eighth flight of March 1966 which had completed 6·5 orbits of the earth. The astronauts Scott and Armstrong are still seated in the spacecraft whilst rescue men assist with the flotation collar before the spacecraft was picked up by the destroyer Mason. There were four further flights in the Gemini project.

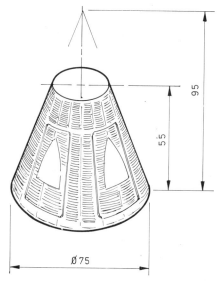

A cardboard model of the spacecraft can be made from the three units shown in Fig. 14.14. Draw orthographic views of the completed model and from them draw a development of each of the three parts.

Fig. 14.15

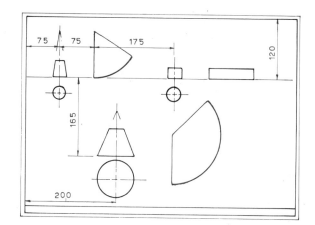

A suggested layout of your drawings is given in Fig. 14.15, although you may wish to use a better layout.

Your model can be finished by adding detail and painting. It is often easier to put surface detail on the development before cutting it out and assembling it. Details drawn with a hard, sharp pencil are usually deep enough to show even when the model is sprayed with silver paint.

Veteran car project

Fig. 14.16

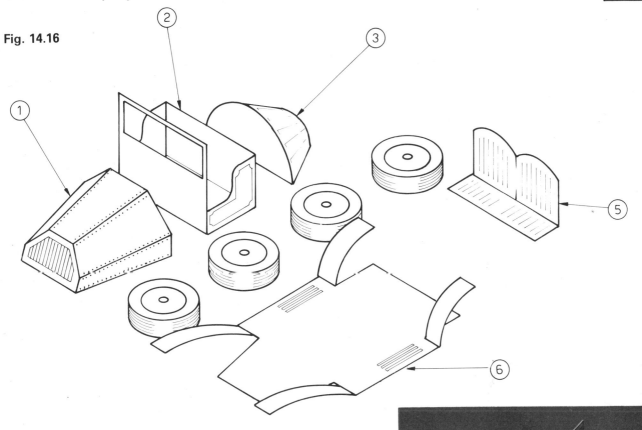

A cut-out cardboard model of a veteran car is shown in break-down form in Fig. 14.16. Details of the component parts of the model are given below and on the following page.

Draw, scale 1:1 in either first or third angle orthographic projection, three views of the assembled model.

Draw sufficient of the given views of the component parts to be able to produce a development of each part.

Transfer the development of each part on to good quality thin card, cut out the parts, adding gluing flaps where necessary and assemble the model. Complete it by painting and adding further detail.

Fig. 14.17

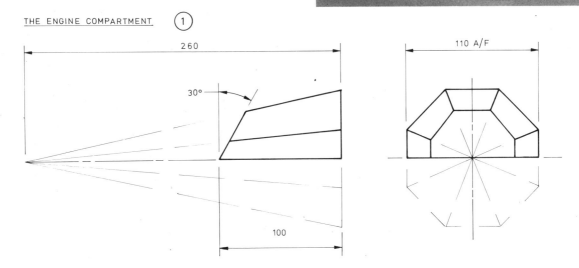

Engine compartment (Fig. 14.17)
This is in the shape of half a truncated octagonal pyramid.

Veteran car project

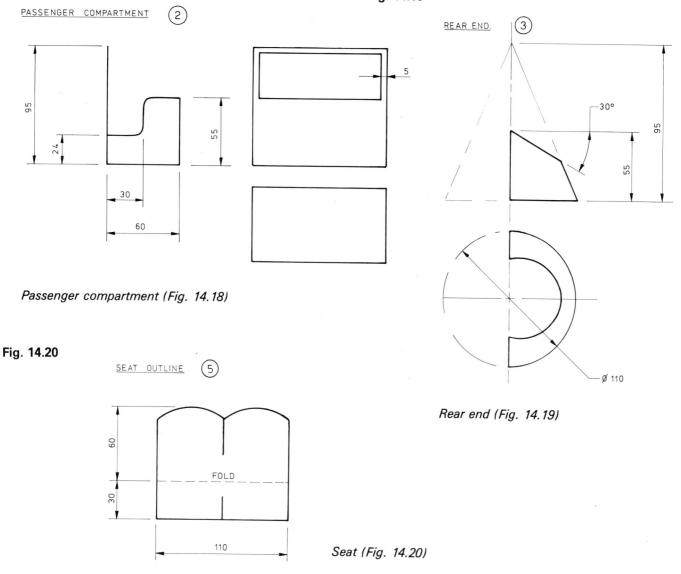

Fig. 14.18

Passenger compartment (Fig. 14.18)

Fig. 14.19

Rear end (Fig. 14.19)

Fig. 14.20

Seat (Fig. 14.20)

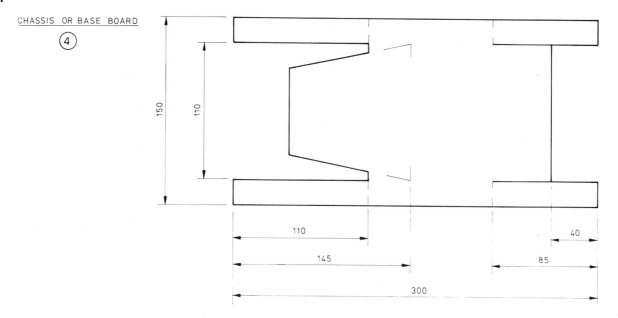

Fig. 14.21

Chassis or base board (Fig. 14.21)
Cut this part from slightly thicker card, and bend the mudguards to the shape shown in Fig. 14.16.

The wheels can be made of cardboard cylinders or turned from wood or plastic.

Toy roller project

An exploded pictorial view of a child's toy road roller made of a hardwood is shown in Fig. 14.22. A detail drawing including a parts list is given in Fig. 14.23.

Draw, scale 1:1, in either first or third angle orthographic projection, the following views of the assembled toy: a front view; an end view; a plan view.

Add to your drawing any design features which you think would improve the appearance, durability and safety of the toy.

Fig. 14.22

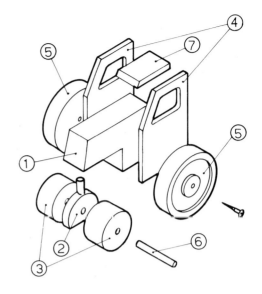

Fig. 14.23

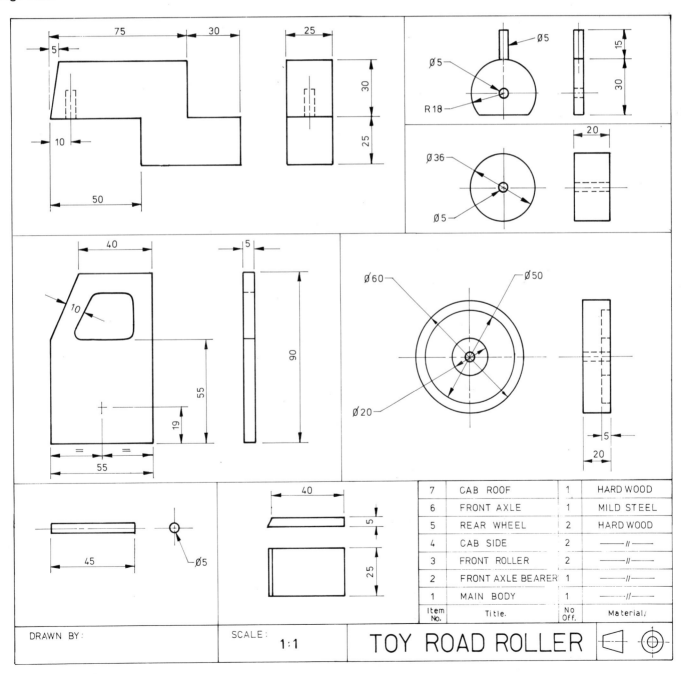

Card modeller's clamp project

The design sketches and notes for producing a card modeller's clamp are shown in Fig. 14.24.
Using this information, draw scale 1:1 in either first or third angle orthographic projection, the following views of the assembled clamp: a front view; an end view; a plan.

Show all hidden detail and completely dimension the drawing.

Fig. 14.24

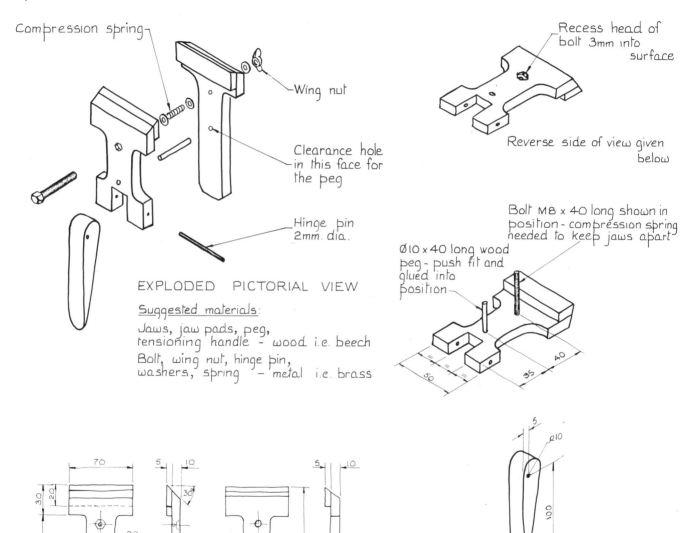

ORTHOGRAPHIC VIEWS OF BOTH JAWS

15 Applied geometry

Intersection – prisms

When two solids penetrate each other, the line at which they meet is called the line of intersection, and it is determined by plotting the projections of points which are common to both surfaces.

When two prisms intersect, the line of intersection between them will consist of straight lines.

An example of the intersection of a triangular prism and a hexagonal prism is given in the orthographic views (Fig. 15.1), and pictorial view (Fig. 15.2). In the orthographic drawing, the end and plan views can be drawn completely, assuming that sizes are given. In the front view, however, the line of intersection has to be plotted.

To illustrate the construction, the edges of the triangular prism have been numbered. The points at which these lines intersect with the hexagonal prism can be projected from the plan and end views to the front view in order to establish the line of intersection as shown.

Fig. 15.1

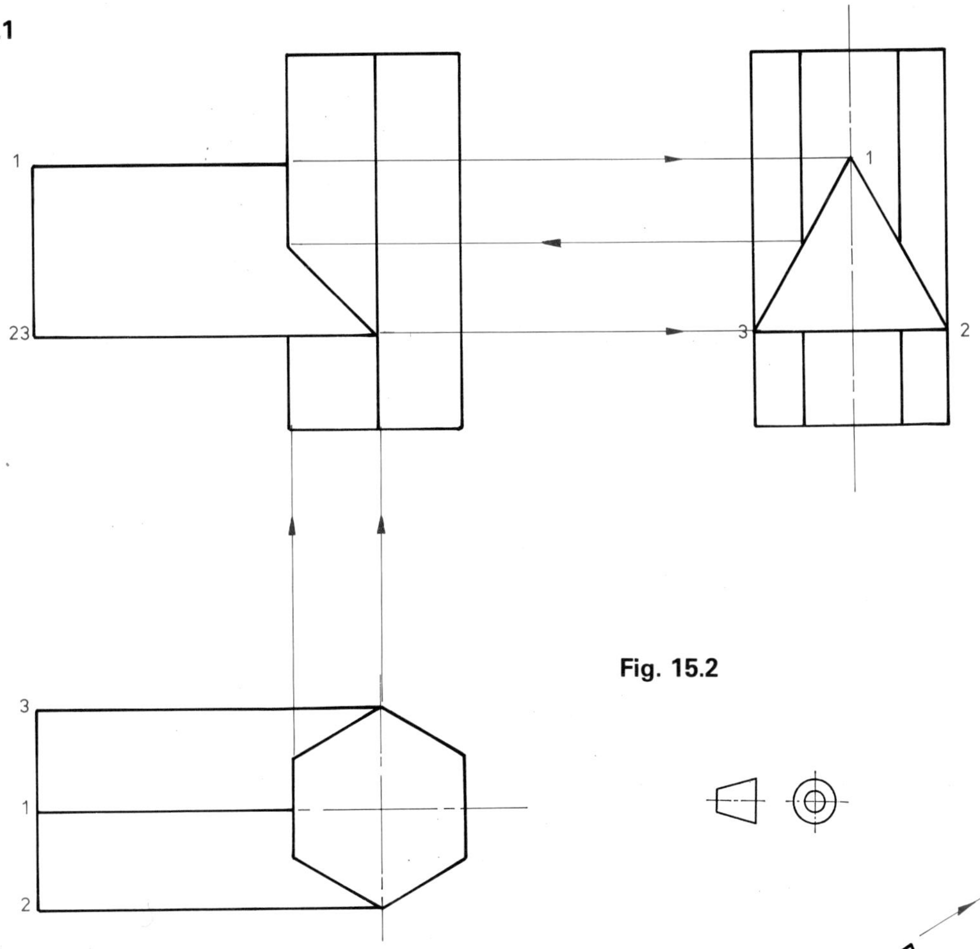

Fig. 15.2

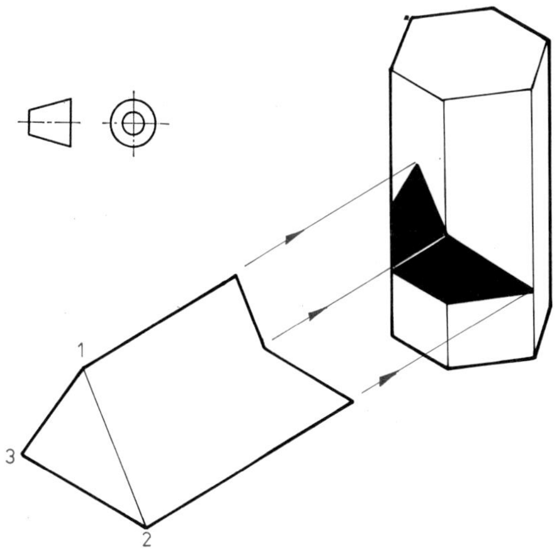

Intersection – examples

Fig. 15.3

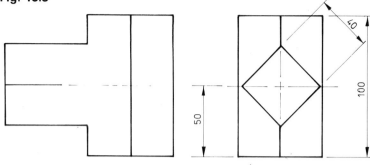
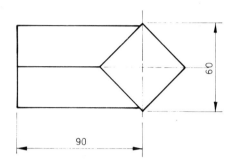

1. Draw, scale 1:1, the three views of the two square-based prisms given in Fig. 15.3, and then add the line of intersection between them.

Fig. 15.4

Fig. 15.5

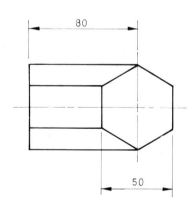
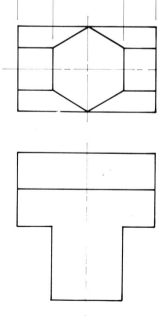
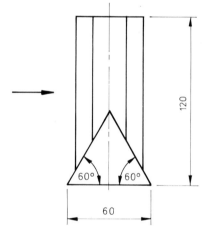

2. Three views of intersecting hexagonal-based prisms are given in Fig. 15.4. Draw the three views scale 1:1, and add the line of intersection in the plan view.

3. Two orthographic views of a triangular-based prism intersected by a hexagonal-based prism are given in Fig. 15.5. Draw the given views scale 1:1, and from them project the end view indicated by the arrow including the line of intersection between the two prisms.

Intersection – cylinders

As there are no edges on a cylinder, compared with the triangular prism on page 75, it is necessary to create artificial edges, called ordinates, on the intersecting cylinder.

Orthographic views of the intersection of two cylinders of unequal diameter are given in Fig. 15.6. A pictorial view of the cylinders is given in Fig. 15.7.

Construction:

Divide the end view of the intersecting cylinder into twelve equal parts.

Number the parts as shown.

Project horizontal lines from these points to the front and plan views.

Number the lines as shown.

Project the points of intersection of these ordinates with the vertical cylinder in the plan view to the similarly numbered lines in the front view.

Join these points of intersection to produce the required line.

Fig. 15.6

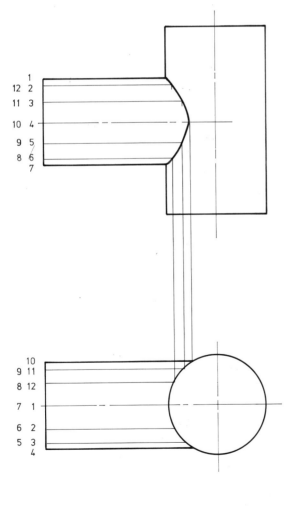

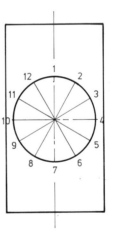

Fig. 15.7

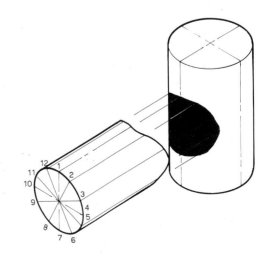

Intersection – examples

Fig. 15.8

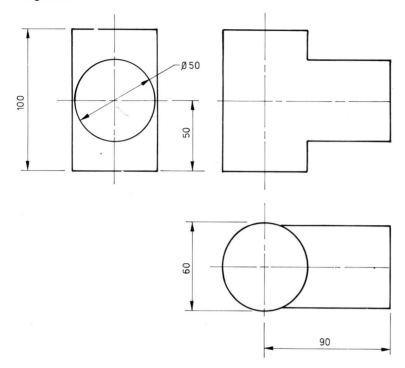

1. The intersection of two cylindrical pipes is shown in the orthographic views Fig. 15.8. Draw the given views scale 1:1 and construct the line of intersection between them.

Fig. 15.9

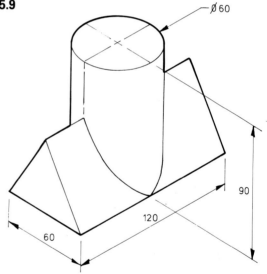

2. A cylindrical tube is to fit over a triangular bar as shown in the pictorial view Fig. 15.9.

   (a) Draw orthographic views showing the line of intersection between the two solids.
   (b) Draw a development of the cylindrical tube.

Fig. 15.10

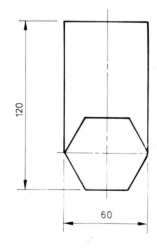

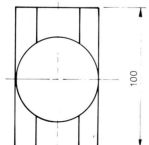

3. Two orthographic views of a hexagonal-based prism intersected by a cylinder are shown in Fig. 15.10. Draw, scale 1:1, the two views and from them project the end view showing the line of intersection between the two solids.

16 Pictorial projection   Oblique

Fig. 16.1   Fig. 16.2   Fig. 16.4

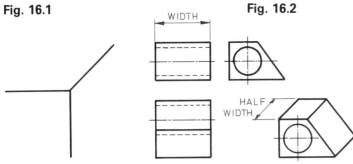

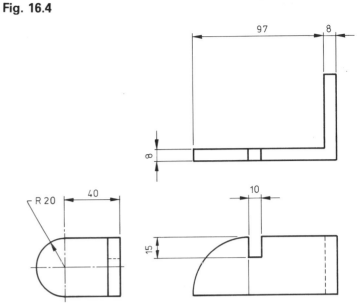

Remember that in oblique projection there is one pictorial view with the axes of the view usually at 90°/135°/135° as shown in Fig. 16.1.

In *Cabinet* oblique, the third axis, i.e. the one showing the thickness of the view is halved as shown in Fig. 16.2.
There is another form of oblique projection, known as *Cavalier* in which the third axis remains full size.

Fig. 16.3

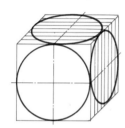

It is an advantage to position objects so that circular faces are drawn to the front. If this is not possible, circles are drawn using the grid method shown in Fig. 16.3.

1 A gate latch, drawn in orthographic projection, is shown in Fig. 16.4.
   Draw the gate latch scale 1:1, in oblique projection.

Fig. 16.5

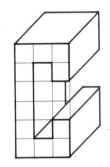

Fig. 16.6

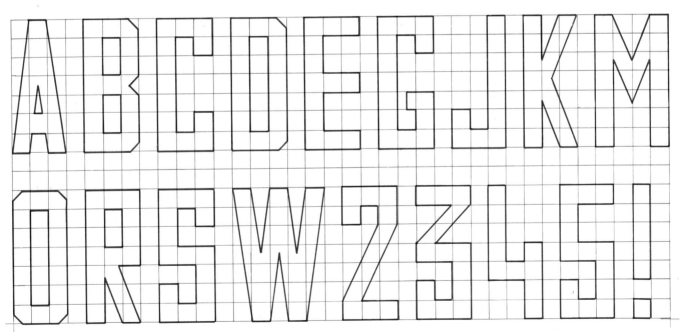

2 Part of an alphabet based on squares is shown in Fig. 16.6. Not all the letters and numerals are given; adapt those given to complete the alphabet.
   Using squares of 10 mm sides, and assuming that the letters are cut out of material 50 mm thick, draw an oblique view of your own initials as in Fig. 16.5.

Can you improve on the shape of the letters? Design an alphabet to be made of plastic suitable for mounting on a shop front. Draw some of the letters in oblique projection in order to visualise their effect.

80 Isometric

In isometric projection there is one pictorial view with the axes of the view at 120°/120°/120° as shown in Fig. 16.7.

Fig. 16.7 **Fig. 16.8**

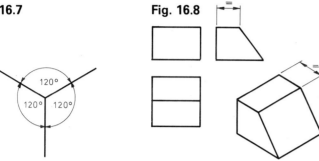

Fig. 16.9

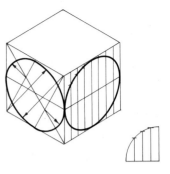

There are two methods for drawing circles and arcs in isometric projection as shown in Fig. 16.9.

Remember always to measure along the three axes (Fig. 16.8).

Fig. 16.11

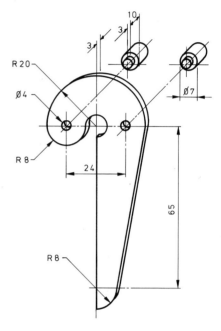

Fig. 16.10

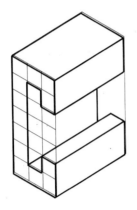

1. Look at Fig. 16.6 on page 79.
Using a grid of squares of 10 mm sides, design and draw a similar alphabet in isometric projection. Fig. 16.10 shows one such letter.

2. An exploded oblique view of a centre square is given in Fig. 16.11.
Draw scale 1:1, an isometric view of the *assembled* centre square.

Fig. 16.12

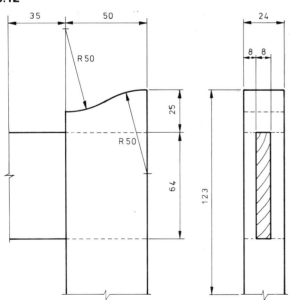

Fig. 16.13

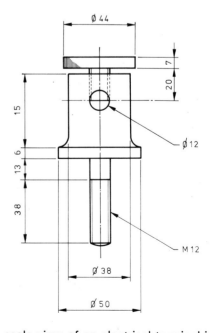

3. Orthographic views of the corner of a wooden gate are given in Fig. 16.12.
Draw an exploded isometric view of the corner.

4. A large-scale view of an electrical terminal is given in Fig. 16.13.
Draw, to the sizes given, an isometric view of the terminal.

Pictorial projection – examples

A4 each

Illustrations of four objects are given in Figs. 16.14, 16.15, 16.16 and 16.17. Choose a suitable pictorial form of projection, and draw each of the objects scale 1:1.

Fig. 16.14

Fig. 16.16

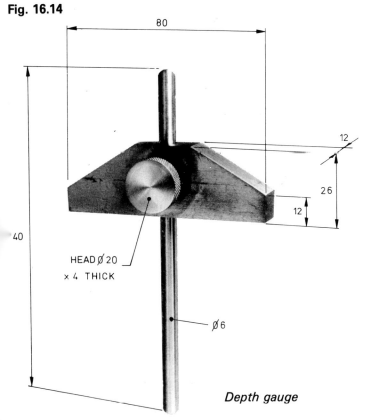

Rocker arm

Fig. 16.15

Fig. 16.17

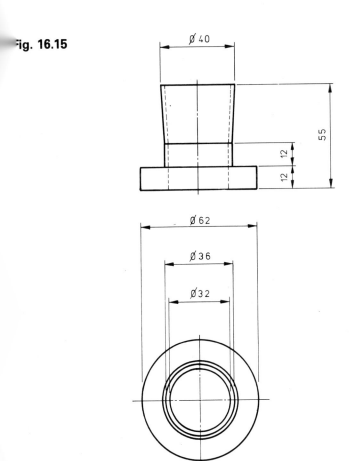

Depth gauge

Batten holder case

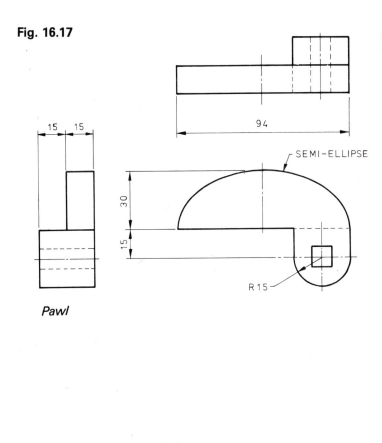

Pawl

82 Pictorial projection – examples

Fig. 16.18

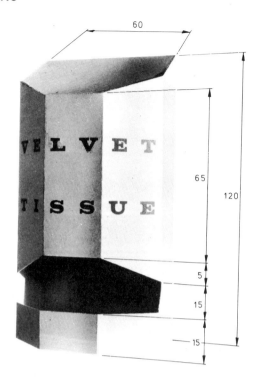

Tissue container

A dimensioned photograph of a cardboard tissue container is given in Fig. 16.18. The container is in the shape of a half of a truncated regular octagonal prism.
Make a scale 1:1 pictorial view of the container.

Fig. 16.19

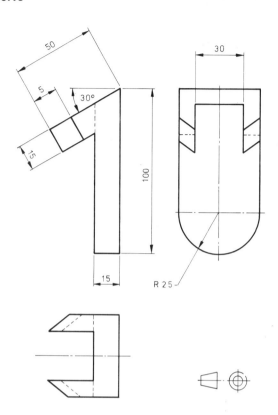

Bracket

Make a scale 1:1 pictorial view of the bracket shown in Fig. 16.19.

Fig. 16.20

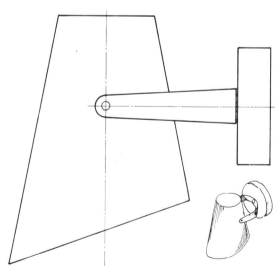

Lamp casing

A view of the casing of a table lamp is given in Fig. 16.20. The shade is in the form of a truncated cone, overall height 150 mm, diameter of top 60 mm, diameter of base 120 mm, and the angle of truncation 15°. The wall-fitting is in the shape of a cylinder, diameter 70 mm, thickness 20 mm, and the bracket a semicircular strip of suitable dimensions. Draw a scale 1:1 pictorial view of the lamp casing.

Fig. 16.21

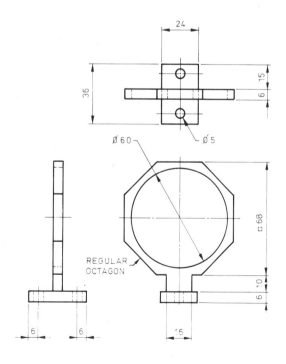

Tumbler holder

Make a scale 1:1 isometric drawing of the tumbler holder shown in Fig. 16.21.

Pictorial projection – examples

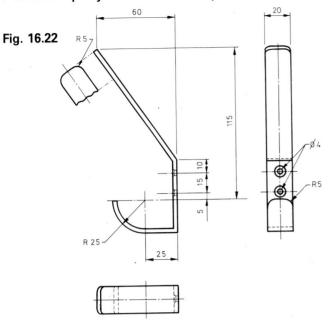

Fig. 16.22

Coat hook

Select a suitable form of pictorial projection and draw a view of the coat hook scale 1:1 (Fig. 16.22). Small radii may be drawn freehand.

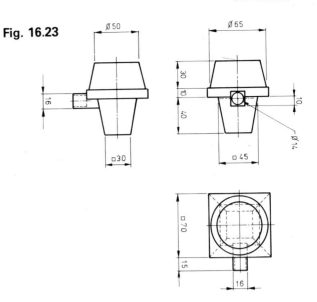

Fig. 16.23

Direction indicator lamp

A direction indicator lamp of a motor cycle, drawn in first angle projection, is shown in Fig. 16.23. Make a full-sized isometric drawing of the lamp. Hidden detail is not required.

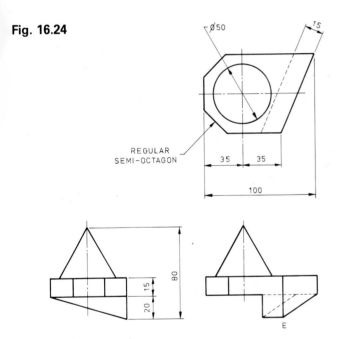

Fig. 16.24

Pivot block

A pivot block, drawn in third angle orthographic projection, is shown in Fig. 16.24.
Draw the pivot block, scale 1:1 and in isometric projection, making the corner *E* the lowest point of the drawing.
Hidden detail is not required.

Kitchen design

A customer proposes to buy the following units of furniture to fit against one wall of the kitchen of her house. The units are designed to be placed together in any order.

To stand on the floor:
(a) a sink unit with cupboards underneath 1200 mm wide;
(b) a set of four drawers 600 mm wide;
(c) two cupboards with a single door each 600 mm wide.

All the above units are 900 mm high and 500 mm from back to front, each has a flat top and a recessed plinth at the base.

To be mounted on the wall:
(d) a cupboard with two doors, 1200 mm wide, 500 mm high and 300 mm from back to front.

A view of the wall on which the units are to be arranged, drawn scale 1:50 is given in Fig. 16.25. The floor-standing units are to be placed in the space 3000 mm long.

In order to help the customer visualise the kitchen as it would be, select a suitable pictorial projection and to scale 1:25, draw a view of the wall including door and window with the units in place.

Estimate any dimensions not given.

Enhance your drawing with colour or shading.

Fig. 16.25

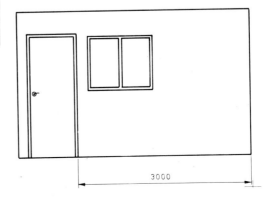

17 Dimensioning

To complete a drawing that is to be used by a craftsman to make an artefact, it is necessary to add to the drawing an indication of the sizes or dimensions of the object in order to define its precise shape and size completely and without ambiguity.

The dimensioning of Engineering Drawing is specified in BS 308 and PD 7308, and that of Building Drawing in BS 1192. These standards apply particularly to orthographic projections and although there are no standards for pictorial drawings, where practicable, the same principles should be applied. Reference to the methods used to dimension pictorial illustrations throughout this book give an indication of the methods that are acceptable. The criteria for dimensioning a drawing must be clarity and unambiguity – in other words, there should be no doubt as to what is meant.

Dimensioning guide

1 *All* essential dimensions must be included. It should not be necessary to scale, i.e. measure, a drawing in order to determine a size.

2 Dimensions should not be repeated on a view or different views.

3 The dimensions shown should be the actual size of the object. Dimensions are not scaled.

4 Wherever possible, dimensions should be placed outside the outline of the view and nearest to the part being dimensioned (Fig. 17.1).

5 Dimension lines should be a minimum of 10 mm away from the outline of the view and, where dimension lines are parallel, at least 10 mm apart. If a pen is used, dimension lines should be between 0·2 and 0·3 mm thick. If a pencil is used, dimension lines should be thinner than the outline.

6 Shorter dimension lines should be nearer the outline with overall dimensions outside.

7 Limit lines, i.e. lines leading from the outline of the object to the dimension line, should start 2 – 3 mm away from the outline, and extend the same distance beyond the dimension line. Limit lines should be the same thickness as dimension lines.

8 Arrow heads should be solid, neat and pointed, and approximately 3 mm long. Arrow heads *must* touch extension lines (Figs. 17.2 and 17.3).

9 Dimension figures should be placed as near as possible above the centre of the dimension line, positioned so that they read either from the bottom or from the right hand side of the drawing.

10 Outlines, centre lines and limit lines must not be used as dimension lines.

11 Hidden detail should not be dimensioned unless it is unavoidable.

12 Circles, i.e. holes, should be dimensioned by the diameter. The abbreviation for diameter is ∅ shown in front of the dimension (Fig. 17.4).

13 Arcs should be dimensioned by the radius. The abbreviation for radius is R shown in front of the dimension (Fig. 17.5).

14 Leader lines for arcs and circles should be in line with the centre.

15 There should be a statement on the drawing indicating the unit of measurement used, e.g. *all dimensions in millimetres.*

Examples of dimensioning are given on the following two pages. Further reference should be made to the appropriate British Standard, copies of which may be obtained from the British Standards Institution, 2 Park Street, London, W1A 2BS.

Dimensioning guide

Fig. 17.1

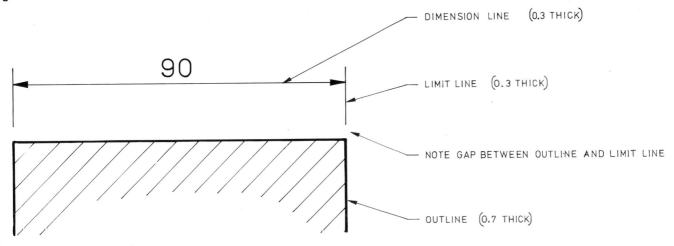

Dimension lines and limit lines

Fig. 17.2

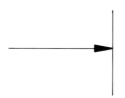

Fig. 17.3

NOT

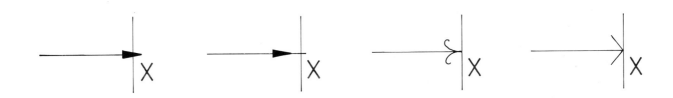

Arrow heads

Fig. 17.4

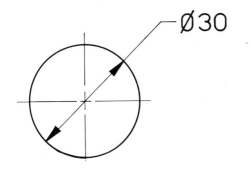

Fig. 17.5

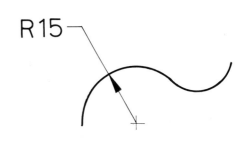

Circles and arcs

86 Dimensioning guide

Fig. 17.6 *Example*

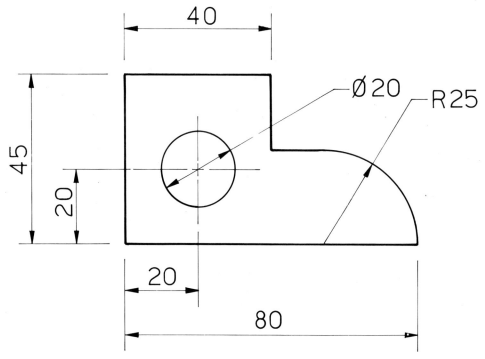

Fig. 17.7 *Small features*

Fig. 17.8

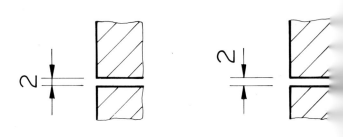

Fig. 17.9 *Angles*

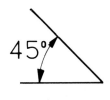

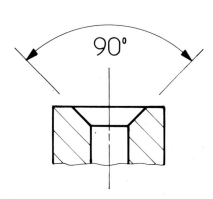

Fig. 17.10 *Concentric holes*

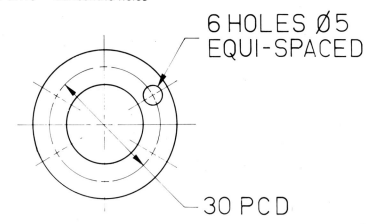

PCD = pitch centre diameter

18 Sections

In orthographic projection, internal detail of an object is represented by broken lines. In complicated objects, the amount of hidden detail can cause confusion. To avoid this, sectional views are drawn: imagine that the object has been cut through and a part removed to expose internal detail. The surface created by the imaginary cut is known as the *section plane;* the part of the object that remains is the section.

It is usual in orthographic projection for sectional views to replace excessive hidden detail.

A simple example of a sectional view is shown in the pictorial illustration Fig. 18.1, and the orthographic views Fig. 18.2.

Note the conventional method for the section line which indicates the position of the section plane. The arrows indicate the direction in which the sectional view is seen.
Note also that the cut surface is cross-hatched. These lines should be drawn at 45° wherever possible, thinner than the outline of the object and 3 – 4 mm apart.

A more complicated example is shown in the pictorial illustration Fig. 18.3, and the orthographic views Fig. 18.4. Notice that only the solid parts of the object, i.e. those that are cut, are cross-hatched.

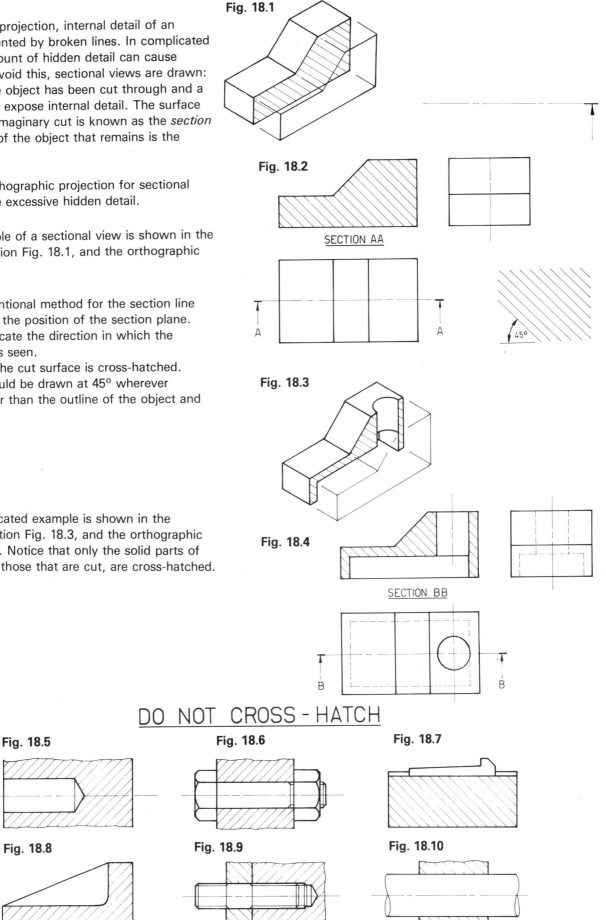

For clarity, some features are not shown as sections even though the section plane passes through them. Examples of these are given in Figs 18.5, 18.6, 18.7, 18.8, 18.9 and 18.10.

Note that in Fig. 18.9 when two adjacent parts are sectioned, the direction of the hatching lines is reversed.

88 Sections – examples

Fig. 18.11

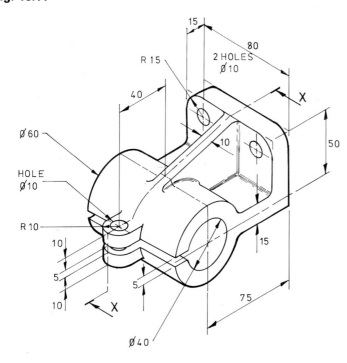

1. A pictorial illustration of a casting is shown in Fig. 18.11.
   Draw, scale 1:1, in first angle orthographic projection:

   (a) the sectional view indicated by the section line *XX*;
   (b) an end view;
   (c) a plan.

   A suggested layout for your drawings is given in Fig. 18.12.

Fig. 18.12

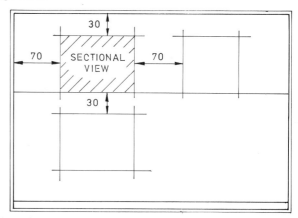

Fig. 18.13

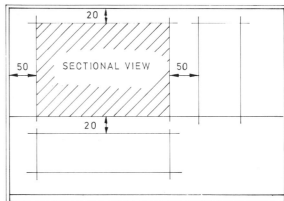

2. A pictorial view of a casting is shown in Fig. 18.13.
   Draw, scale 1:1, in first or third angle orthographic projection, three views of the casting of which one is to be the sectional view *FF*.

   A suggested layout for a first angle drawing is given in Fig. 18.14.

Fig. 18.14

19 Screw threads

Fig. 19.1

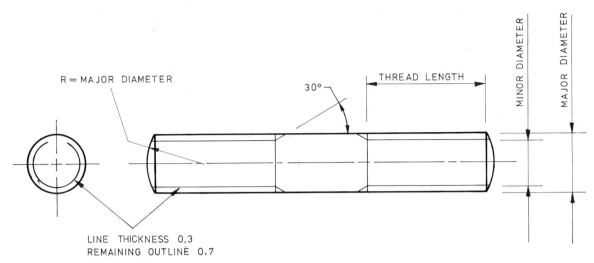

Because it would be time-consuming to draw actual screw threads, standard conventions are used. Details of these are given in BS 308 and PD 7308 which can be obtained from the British Standards Institution, 2 Park Street, London W1A 2BS.

External screw threads (Fig. 19.1)

The minor diameter of the thread is indicated by two thin lines parallel to the major diameter in the side view, and by a broken line in the circular end view. The run-out, showing the termination of the thread along the shank, is at 30° as shown.
The minor diameter is drawn 0·8 of the major diameter. Line thicknesses are as indicated.

Remember that bolts and studs are not sectioned.

Fig. 19.2

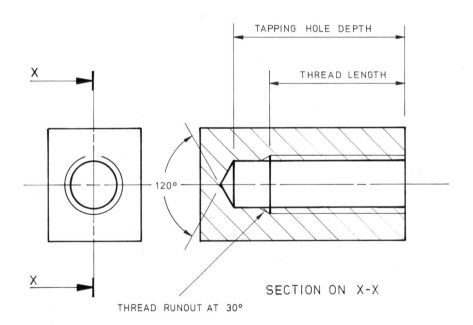

Internal screw threads (Fig. 19.2)

The tapping hole is the same diameter as the minor diameter in Fig. 19.1. When sectioning a threaded hole, the hatching lines must be extended to the tapping hole as shown.

Screw threads

Given below are the steps for drawing a standard nut and bolt. It is quicker and easier to use a radius aid for drawing the curves, or to use a complete template, and these should be used whenever possible.

As a general rule, always show a hexagonal nut or bolt head with the three faces presented in both side and end views.

Fig. 19.3

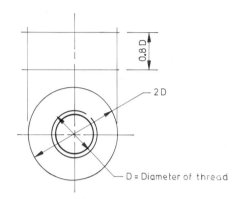

Fig. 19.4

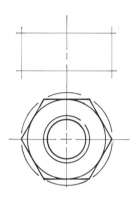

Fig. 19.5

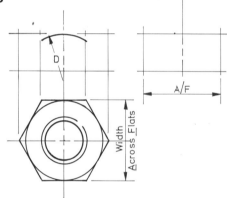

Fig. 19.6

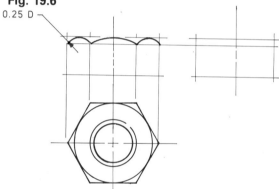

Fig. 19.7

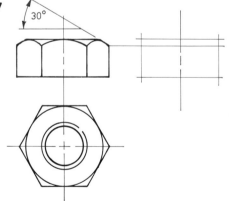

Fig. 19.8

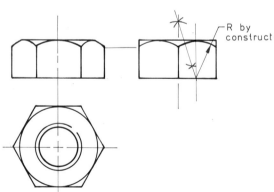

Fig. 19.9

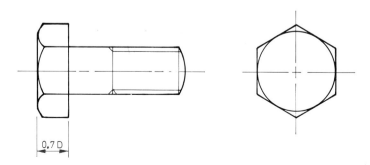

20 Engineering graphics

Power take-off casting

A pictorial view of a power take-off casting is given in Fig. 20.1. Draw, scale 1:1 in first or third angle orthographic projection, the following views of the casting:

(a) the sectional front view, the plane of the section being indicated by the section line *XX*;
(b) a plan view;
(c) an end view.

Hidden detail should be included in the plan and end views only. Dimensions not given are left to your discretion.

Fig. 20.1

Pump cover

A part sectioned pictorial view, together with an orthographic view of the face of a pump cover are given in Fig. 20.2.
Draw, scale 1:1, in either first or third angle orthographic projection, the following views of the cover:

(a) a full sectional front view taken on the principal vertical centre line;
(b) an end view;
(c) a complete plan.

Hidden detail should be included in the most appropriate view.

Fig. 20.2

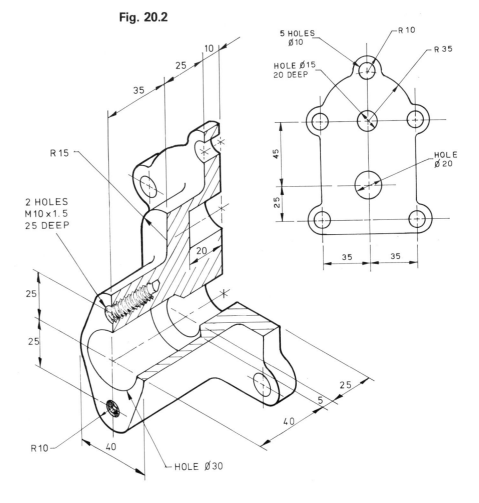

Engineering graphics – examples

1 Corner steady

A dimensional pictorial drawing of the body casting and tommy bar of a corner steady are given in Fig. 20.3, together with a scrap view of the complete assembly. The corner steady is retained in position on a Ø 40 rod by the pressure of the threaded tommy bar.

Draw, scale 1:1, in either first or third angle orthographic projection, three views of the assembled corner steady in position in the middle of a Ø 40 vertical rod 120 mm tall. Select three views that give the greatest amount of information. Include hidden detail.

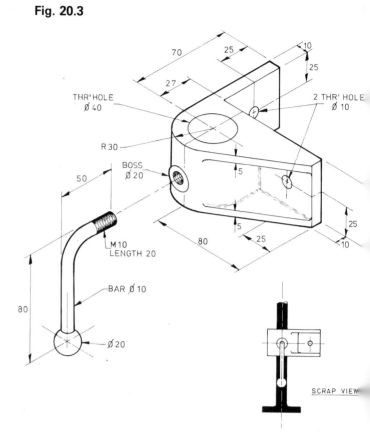

Fig. 20.3

2 Winding handle

An exploded view of a winding handle assembly together with a scrap view of the handle and retaining bolt are given in Fig. 20.4. The ribbed casting is prevented from turning on the shaft by a square feather key and is retained by a M 20 nut and spring washer. The handle is attached to the other end of the casting by a slotted round-headed bolt.

Details of the assembly and dimensions are given in the scrap view.

Draw, completely assembled and in first or third angle orthographic projection, three views of the winding handle of which one must be a sectional view. The sectional view should be the one which gives the maximum detail of construction.

Fig. 20.4

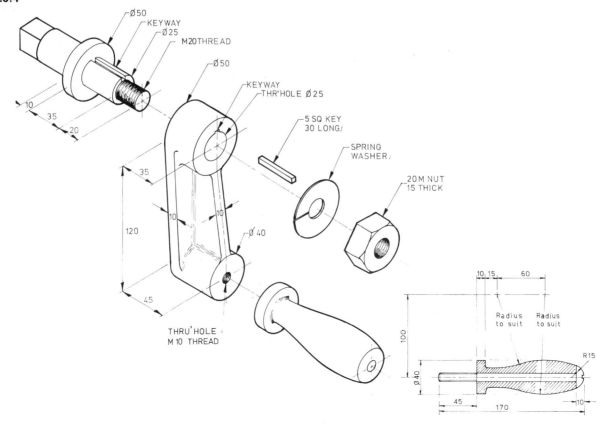

Engineering graphics — examples

Fig. 20.5

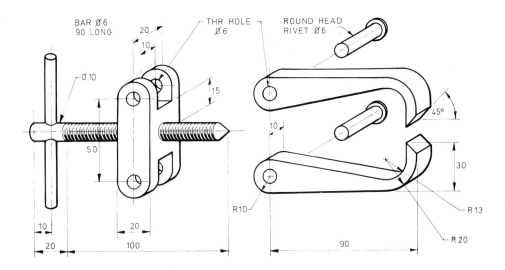

3 Draw bolt

An exploded view of a draw bolt is shown in Fig. 20.5. The two claws are held in the yoke by round-headed rivets and the threaded bar screws into the centre of the yoke.

Draw, scale 1:1, in either first or third angle orthographic projection, three views of the assembled draw bolt of which one must be a sectional view.

4 Strop assembly

A pictorial view of a strop assembly is given in Fig. 20.6. It consists of a body in which has been cut a slotted hole, and underneath, obscured from view in the pictorial illustration but shown in the scrap view, is an elongated hole.

The square nut moves along the slotted hole while the threaded bar is screwed through the nut and out through the elongated hole. At the bottom end of the bar is attached a circular shoe which is retained in position by a C clip located in a groove.

Draw, scale 2:1, in either first or third angle orthographic projection, the following views of the assembled tool:

(a) a front view looking directly on the inverted U-shape of the body;
(b) a sectional end view taken through the centre of the bar;
(c) a plan.

Fig. 20.6

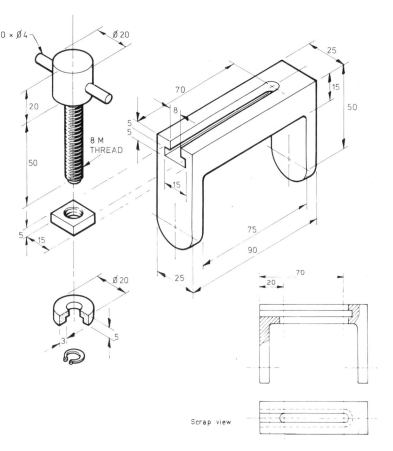

5 Support bracket

An exploded pictorial view of a wall-mounted support bracket is given in Fig. 20.7. It consists of the main body into which the two studs are threaded, and a cap located over the studs which is retained in position by two hexagonal nuts. (The two nuts are not shown in the pictorial view.)

Draw, scale 1:1, in either first or third angle orthographic projection, the following views of the assembled unit:

(a) a sectional front view taken on the centre line of the three holes;
(b) an end view looking from the left of the front view (a);
(c) a plan.

Complete your drawing by adding hidden detail to the plan view only, and by printing the title, scale and symbol of projection used in a suitable position.

Fig. 20.7

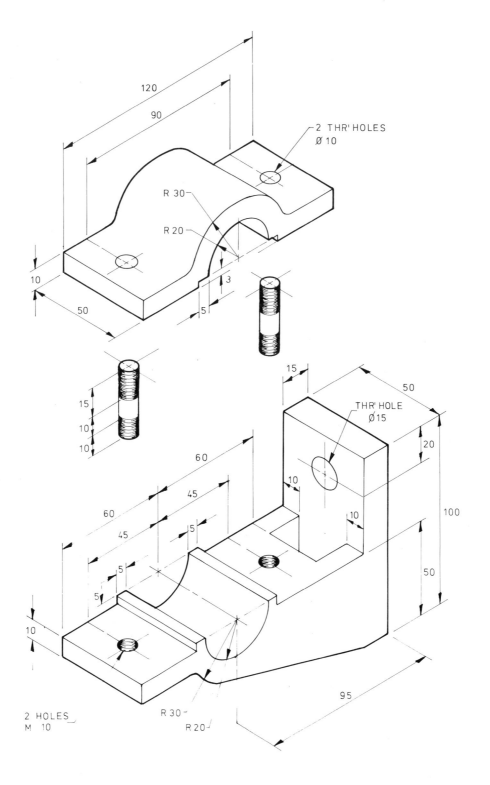

Engineering graphics – examples

6 *Adjustable spanner*

A pictorial view of an adjustable spanner is given in Fig. 20.8. Orthographic views of the component parts of the spanner are given in Fig. 20.9. Divide the A2 sheet of paper into two equal parts by drawing a vertical line.

In one half of the sheet, draw, scale 1:1, in either first or third angle orthographic projection, three views of the assembled spanner with the jaws opened to the full extent.
In the other half of the paper, draw, scale 1:1, an exploded isometric view of the spanner. Enhance this drawing with shading and/or colour.

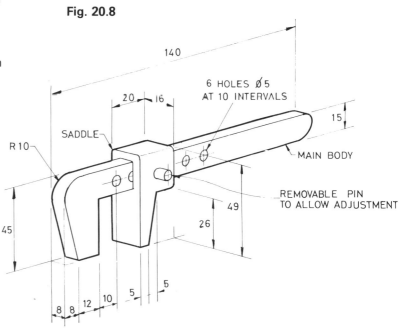

Fig. 20.8

Fig. 20.9

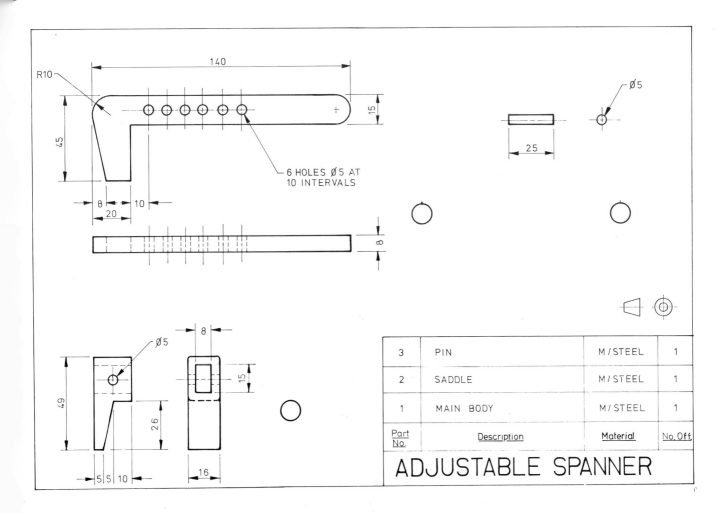

Engineering graphics – examples

7 *Sash cramp*

Detail drawings of the components which make up the adjustable jaw of a sash cramp are given in Fig. 20.10, together with a photograph of the assembly (Fig. 20.11).

The fixed end-piece is attached to a T-bar by two rivets. The square threaded bar passes through the fixed end-piece and into the sliding jaw. The threaded bar is retained in position by the spacing collar and pin.

Draw, scale 1:1, in either first or third angle orthographic projection, the following views of the components assembled on a suitable length of T-bar:
(a) a sectional front elevation, the section plane being taken on the principal centre line;
(b) an end elevation;
(c) a plan.

Do not section the T-bar in the sectional view.

Fig. 20.10

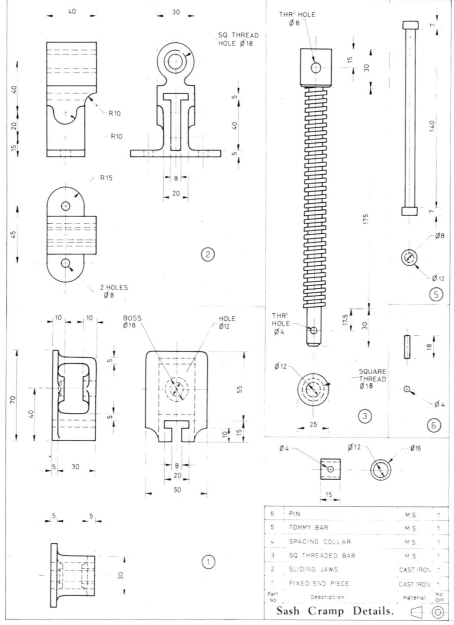

Fig. 20.11

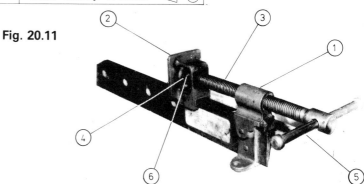